Bondability of modified wood

AF307761

Bondability of modified wood

Dissertation

In Partial Fulfillment of the Requirements

for the Doctoral Degree (Dr. rer. nat.)

of the Faculty of Forest Science and Forest Ecology

Georg-August-University Götttingen

Submitted by

Alireza Bastani

Born in Ahvaz, Iran

Göttingen, 2016

Bibliografische Information der Deutschen Nationalbibliothek

Die Deutsche Nationalbibliothek verzeichnet diese Publikation in der

Deutschen Nationalbibliografie; detaillierte bibliografische Daten sind im Internet

über http://dnb.d-nb.de abrufbar.

1. Aufl. - Göttingen: Cuvillier, 2016

  Zugl.: Göttingen, Univ., Diss., 2016

Supervisors:

Prof. Dr. Holger Militz

Prof. Dr. Stergios Adamopoulos

Date of disputation: 23.03.2016

  ISBN 978-3-7369-9289-4

  eISBN 978-3-7369-8289-5

Acknowledgments

First and foremost, I would like to express my sincere gratitude to Prof. Dr. Holger Militz, not only for supervising this research but also for his endless and kind supports during my PhD work.

Great and sincere thanks to Prof. Dr. Stergios Adamopoulos also for supervising this research. His valuable advices are much appreciated.

I would also like to thank all of my colleagues in department of wood biology and wood products in Göttingen University for their friendship and cooperation.

Cordial thanks to my parent and sisters for their continuous encouragement and to my wife for supporting me in my life during both good and difficult times.

At the end, I would like to dedicate this work to my dear son "Sam". I hope this will lead him to a very bright future.

Table of content

Publications 1-7

1.   Adamopoulos S, Bastani A, Gascon-Garrido P, Militz H, Mai C (2012) Adhesive bonding of beech wood modified with a phenol formaldehyde compound. Eur J Wood Prod. 70:897–901

2.   Bastani A, Adamopoulos S, Militz H (2015) Water uptake and wetting behaviour of furfurylated, N-methylol melamine modified and heat-treated wood. Eur J Wood Prod. 73(5):627-634

3.   Bastani A, Adamopoulos S, Militz H (2015) Gross adhesive penetration in furfurylated, N-methylol melamine modified and heat-treated wood examined by fluorescence microscopy. Eur. J. Wood Prod. 73:635–642

4.   Bastani A, Adamopoulos S, Koddenberg T, Militz H (2015) Study of adhesive bondlines in modified wood with fluorescence microscopy and X-ray micro-computed tomography. International Journal of Adhesion Adhesives. 68(2016)351–358

5.   Bastani A, Adamopoulos S, Militz H (2015) Shear strength of furfurylated, N-methylol melamine and thermally modified wood bonded with three conventional adhesives. Wood Material Science and Engineering. DOI:10.1080/17480272.2016.1164754

6.   Bastani A, Adamopoulos S, Militz H (2015) Effect of open assembly time and equilibrium moisture content on the penetration of polyurethane adhesive into thermally modified wood. Journal of Adhesion. DOI:10.1080/00218464.2015.1118621

7.   Bastani A, Adamopoulos S, Rohumaa A, Militz H (2015) Development of bonding strength of modified birch veneers during adhesive curing. Wood Research. 61 (2): 2016 205-214

1. Summary

This study investigates the bonding properties of modified wood by considering three different aspects: water related characteristics, mechanical performance and optical (fluorescence microscopy and X-ray micro-computed tomography) observation of adhesive penetration into modified wood structure. In recent years, the new wood modifications have become more commercially available in the market for both exterior and interior applications due to improved properties that modification can bring to the wood e.g. the improved biological durability, dimensional stability, hardness and weathering resistance of the wood as well as the environmentally friendly nature of the wood modification processes (Militz and Hill 2005). Besides these advantages, modification can affect some technological aspects of the wood such as its bonding performance. For example, it can alter the strength of adhesion as a result of changes in chemical, physical and structural characteristics of the wood. For example, the less polar and less porous modified wood surfaces can result in reduced adhesion due to formation of less free -OH groups for bonding leading to poorer adhesive wetting of the wood surface and weaker chemical bonds between the two adherents (Hunt et al. 2007). As modified wood becomes a more demanded material for different applications, there is a need to study its bonding performance where the challenge is to bond different modified materials as their physical and chemical characteristics are substantially changed by modification.

In this thesis, measurements of capillary water uptake, contact angle and surface energy were used to determine the water related properties and hydrophobic behavior of furfurylated (FA40 and FA70, which represent 65 and 75 % WPGs) and N-methylol melamine (NMM) (10, 20 and 30%) modified Scots pine and thermally treated Scots pine and beech (modified through an industrial scale vacuum press dewatering method at 195 and 210 °C). The capillary water uptake results indicated a considerable reduction of water uptake for all modifications in all directions both after short (24 h) and long contact times (168, 336 h). Contact angle measurement data revealed an increased hydrophobicity of modified wood. However, some exceptions were observed, mainly for thermally -treated wood. Modifications provided radial and tangential surfaces with a non-polar character. Penetration of adhesives into the wood structure plays an important role in the production of glued wood-based panels and products by affecting the bond quality (Frihart 2005, Kamke and Lee 2007). The gross penetration of emulsion polymer isocyanate (EPI), polyurethane (PU) and polyvinyl acetate (PVAc) adhesives into modified wood, both with and without pressure, were determined by using fluorescence microscopy based on measurements of effective (EP) and maximum penetration (MP). Without application of pressure, the EP of EPI adhesive reduced after NMM modification and furfurylation (FA70) and also PU adhesive after NMM modification while the EP of PVAc adhesive increased into furfurylated and NMM modified (10 and 20%) wood. For thermally treated Scots pine, increasing the treatment temperature improved EP of all adhesives. Among used adhesives, PU penetrated much deeper into thermally -treated wood for both treatment temperatures. Comparison of penetration of adhesive with and without pressure revealed that with the exception of EP of PU and EPI adhesives into NMM-modified wood and PVAc into thermally treated beech at 195°C, application of pressure led to rather different results as compared to the EP data when no pressure was applied. Visual observation and analysis of fluorescence microscopy photomicrographs provided more detailed information on modality of penetration. Due to the large and deep penetration of PU adhesive into thermally treated Scots pine observed in both studies (with and without pressure), the 3D pattern of penetration of this adhesive was obtained by X-ray micro-computed tomography indicating the pathways which were used by this adhesive for penetration. In another study, the bonding shear strength of the same modified wood materials glued with the same adhesives was also investigated. For all adhesives used, the shear strength significantly reduced after furfurylation and NMM modification of Scots pine samples, mainly due to the brittle nature of the wood after modification rather to the failure of the bondline. Bonding strength of both Scots pine and beech was also negatively affected by thermal modification and the bondline was found to be the weakest link in thermally modified wood. The EP of adhesives and the bondline thickness did not relate to the shear strength of all modified wood materials. It was indicated that the lower shear strength of modified wood could be attributed to other factors, such as the decreased chemical bonding or mechanical interlocking of adhesives, and the reduced strength of brittle modified wood substrate.

The effect of two important bonding variables, wood moisture content and open assembly time on penetration of PU adhesive into thermally modified wood (195 and 210 °C) was also studied. The equilibrium moisture content (EMC) level of 8.6% was found to be the optimum for an effective penetration of PU adhesive in thermally modified Scots pine treated at 195°C. In most of the cases, penetration of PU adhesive did not change significantly by increasing the open assembly time, which suggested using a shorter open assembly time of 15 min than 30 min for bonding of thermally modified Scots pine with PU adhesive, in order to save time and reducing the production costs. For samples treated at both treatment temperatures and after shorter open assembly time, the highest MP values observed at moderate EMC levels of 8.6 and 8.2% and the lowest at the higher EMC levels of 13.2 and 12.5%.

In another study, the effect of phenol formaldehyde (PF) treatment on bonding performance of beech glued with PVAc and phenol resorcinol formaldehyde (PRF) adhesives was also investigated. The results of both dry and wet conditions indicated higher shear strength for samples bonded with PRF than PVAc. With the exception of 25% PF treated wood bonded with PVAc, the PF modified wood can be glued with both adhesives satisfactorily under dry condition, while under wet condition only the 25% PF modified samples bonded with PRF provided acceptable bonding. For both adhesive systems, PF modification caused a reduction of adhesive penetration into wood structure, especially in the case of higher load treatment.

The development of bonding strength of modified birch veneers glued with hot curing phenol formaldehyde (PF) adhesive was investigated in different pressing (20 s , 160s) and open assembly times (20s , 10 min). Generally, the bonding strength improved by extending the pressing time. In 20 s pressing, increasing assembly time did not change the bonding strength in most of the cases while at 160 s pressing, prolongation of assembly time developed a better bonding for controls, NMM modified and thermally treated veneers at 180°C. The combination of 10 min assembly time and 160 s pressing time provided the highest bonding strength for controls, NMM modified and thermally treated veneers at 180°C while furfurylated samples achieved the highest values in 20 s assembly and 160 s pressing times. In general, modification affected negatively the bonding performance of the veneers, especially for furfurylated and NMM modified samples. In General, the overall results obtained in this thesis showed that modified wood has lower bonding ability and performance than unmodified wood as result of the decreased water related properties, less penetration of adhesive into wood structure and decreased bonding strength after modification. However, the increased dimensional stability and low water uptake of modified wood might lead to better performance in long term.

2. Introduction

2.1. Wood and adhesion

2.1.1. Wood; a natural, renewable raw material

Wood is the most frequently used building material in the world for centuries. It is a widely distributed, renewable and multifunctional substance, with a high aesthetic value, strong owing to its high strength to weight ratio and easy to process. Wood is a porous and anisotropic material with various anatomical features. It is varied in species, characteristics and uses. Due to wide range of wood applications, it can be found in our modern living environment almost everywhere and it can be produced and used in a sustainable manner according to its renewable nature. Wood products are made using tools, nails and screws and also by cutting wood into smaller size components and re-joining by adhesives or combining with other materials. They can be ranged from simple handmade wooden furniture to highly engineered wood products manufactured in a highly automated production site.

2.1.2. Furniture and building constructions

Among different wood products, the glued wood products have an important share in furniture and construction markets and adhesive bonding of the wood is known as the most important joining technique to connect wooden elements to each other in modern wood construction and timber engineering (Kamke and Lee 2007). A quick look to the vast applications of the glued wood products in interior and exterior decoration e.g. in production of facades, wood partition walls, cabinets and in production of the home and office furniture, chairs and tables, closets, library shelves and in building constructions e.g. structural wood columns, prefabricated walls, stairs, fencing, decking and flooring, attics, garages, window frames, doors, kitchen design, and isolation boards indicates the importance of the adhesion and bonding matter of the wood.

2.1.3. Adhesion background and theories

Adhesion is defined as the situation, in which two surfaces of adherents are kept together by interfacial forces e.g. valence forces, interlocking action, or both (Vick 1999). The history of adhesive bonding dates back to early mankind and Egyptians were cited as the first mankind who bonded wood (Skeist and Miron 1990, River 1994, Keimel 2003, Frihart 2005). There are several mechanisms involved in adhesion such as mechanical interlocking, covalent bonding and secondary interactions e.g. hydrogen bonds and the process of adhesive bond formation in the wood is divided into flow, transfer, penetration, wetting, and solidification steps. The flow is described as spreading of the liquid adhesive along the external wood surface. The transfer refers to the movement of the liquid adhesive to the next wood surface during assembly time and penetration happens due to capillary forces within the cell lumens and bulk flow as result of applied pressure. Wetting, the relative ability of liquid glue to perform interfacial affinity for an adherent, applies on wood surface and causes a uniform flow of adhesive along the walls of the cell lumens. In the last step, which is called solidification, the adhesive transforms into a rigid polymer through polymerization process (Marra 1992, Sernek et al. 1999). As suggested by Marra (1992), an adhesive bond can be also supposed as a chain-link including nine attached links where the maximum strength of the bondline is determined by the weakest link of the chain (Figure 1). According to this plot, link 1 represents the pure adhesive part which is located in middle of the bondline, links 2-3 exhibit the nonhomogeneous and partly cured boundary layer of adhesive, and links 4-5 stand between this boundary layer and the wood material. These later two links represent the adhesion mechanism, e.g. mechanical interlocking, covalent bonding, or secondary chemical bonds as mentioned before. Links 6-7 represent the wood substrate affected by the wood surface preparation or bonding processes, and links 8-9 represent the non-interacted wooden material. Penetration of adhesive affects links 4 through 7, and consequently all possible adhesion mechanisms. The volume including both wood cells and adhesive is called interphase region and the two adherent substrates, each with its own interphase, and the interface between the substrates, form the bondline (Kamke and Lee 2007).

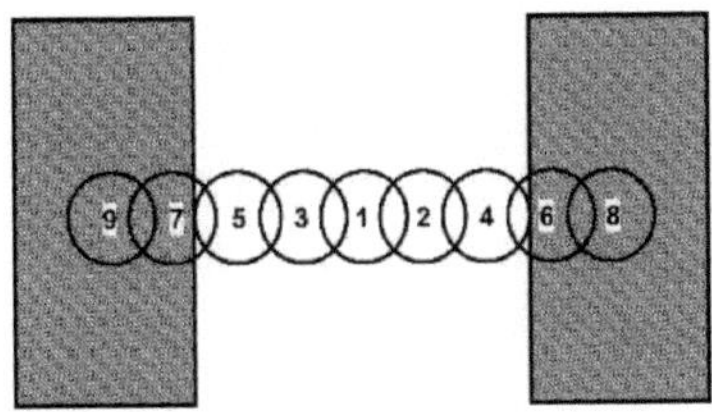

Figure 1: Chain link plot suggested by Marra for an adhesive bond in wood (Kamke and Lee 2007)

Table 1 exhibit the accessibility of the main wood constituents, cellulose, lignin, and hemicellulose to the adhesive for bonding and their involvement to the hydrogen bonding. As can be seen in this table, although hemicellulose is a less abundant constituent of the cell wall, it is quite accessible for bonding and hence it has more contribution than cellulose and lignin.

Table 1: Accessibility of the main wood constituents and their involvement to hydrogen bonding

	Constitution of cell wall (%)	Accessibility for bonding (%)	Involvement in bonding (%)
Cellulose	50	35	17.5
Hemicellulose	24	100	24
Lignin	26	30	7.8

(Salehuddin 1970, River 1991)

2.1.4. Influencing factors on bonding

The influencing factors on bond quality are mainly classified in three groups: wood (e.g. species, cutting direction, free surface energy and moisture content of the wood), adhesive (e.g. type, formulation, molecular weight and viscosity) and process related factors (e.g. open assembly time, pressing time, pressure and temperature). These bonding variables have an interplay effect on each other, which determines the final cost of production and the ultimate bonding quality. For example in many cases, higher load or temperature of pressing might lead to higher bonding quality which can also increase the production costs. Therefore, it is essential to have an optimal combination of these bonding factors, both to meet customer's needs by increasing the quality of the product and also to decrease production costs. Service related factors can be also considered as another group of bonding variables. More detailed information are given in Table 2 (Frihart 2005).

Table 2: Influencing factors on bonding

Resin	Wood	Process	Service
Type	Species	Adhesive amount	Strength
Viscosity	Density	Adhesive distribution	Shear modulus
Molecular weight	Moisture content	Relative humidity	Swell—shrink resistance
Mole ratio of reactants	Plane of cut: radial, tangential,	Temperature	Creep
Cure rate	Heartwood vs. sapwood	Open assembly time	Percentage of wood failure
Total solids	Juvenile vs. mature wood	Closed assembly time	Failure type
Catalyst	Earlywood vs. latewood	Pressure	Dry vs. wet
Mixing	Reaction wood	Adhesive penetration	Modulus of elasticity
Tack	Grain angle	Gas-through	Temperature
Filler	Porosity	Press time	Hydrolysis resistance
Solvent system	Surface roughness	Pretreatments	Heat resistance
Age	Drying damage	Posttreatments	Biological resistance:
	Machining damage		fungi, bacteria, insects,
pH	Dirt, contaminants	Adherend temperature	Finishing
Buffering	Extractives		Ultraviolet resistance
	pH		marine organisms
	Buffering capacity		
	Chemical surface		

(Frihart 2005)

2.1.5. Evaluation methods

Bonding performance of the wood can be examined considering different aspects e.g. by investigation of the physical, chemical and water related properties of the surface of wooden adherent, via mechanical testing of the adhesive joints and glued assemblies and by optical observation of the bondline and adhesive penetration into wood structure. In each category, various methods can be employed to evaluate bonding properties. The physical and chemical conditions of the wood surface are very important to reach a satisfactory bonding performance since adhesives join to the wood by surface attachment. Polarity, surface energy and the amount of extractives on the wood surface have influence on wettability of the wood surface and hence its bonding (Nussbaum 1999, Vick 1999). For example, according to the wetting or adsorption theory the less polar and in some cases less porous wood surfaces can lead to reduced adhesion as result of poorer adhesive wetting of the wood (Hunt et al. 2007). The wettability of wood can be determined through measurement of the contact angle, surface free energy and work of adhesion using parameters which describe the molecular polar or non-polar interactions between liquids and solids (Mantanis and Young 1997, de Meijer et al. 2000, Wålinder and Bryne 2006). Contact angle measurement is a useful method to appraise the solid-liquid interfacial forces. It can be performed by using the Wilhelmy, the rising height, and the sessile drop methods (Pétrissans et al. 2003, Bryne and Wålinder 2010). Estimation of the water uptake and water vapor sorption ability of the wood is another useful method to predict the similar uptake of the liquid adhesive by the wood tissue due to its porous nature and capillary forces. Using different methods, water uptake and water vapour sorption properties of the wood have been investigated in several studies (Donath et al. 2006, Ghosh et al. 2009, Scholz et al. 2009, Johansson and Kifetew 2010, Xiao at al. 2010, Xie at al. 2010, Ghosh at al. 2013, Pries at al. 2013).

Generally, it is expected that an adhesive should be able to keep materials together and to transfer the loads from one adherent to the other one. Mechanical testing of the bonded joints and assemblies are used in order to predict the performance of wooden structures under long term stress and can be divided into four major stressing modes; shear, tensile, cleavage, and peel (Vick 1999). Among mentioned modes, shear mode was taken into consideration in several studies and usually is expressed along with the amount of wood failure of the bondline (Vick and Rowell 1990, Frihart et al. 2004, Brandon et al.2005, van der Zee et al. 2007, Kurt et al. 2008, Sahin Kol et al. 2009).

The wood–adhesive interface, bondline and adhesive penetration can be monitored in several ways e.g. by optical microscopy of cross-sections and micro-slides (Johnson and Kamke 1992, Sernek et al. 1999,), scanning electron microscopy (Koran and Vasishth 1972, Saiki 1984), energy-dispersive X-ray analysis (Smith and Côté 1971, Bolton et al. 1988), scanning thermal microscopy (Konnerth et al. 2008), UV-microscopy (Gindl et al. 2002), energy loss spectroscopy

(Rapp et al. 1999), confocal laser scanning microscopy (Xing et al. 2005), neutron radiography (Niemz et al. 2004), and X-ray tomography to have a three-dimensional view of wood tissues (Shaler et al. 1998; Modzel et al. 2011; Hass et al. 2012; Paris et al. 2013; Kamke et al. 2014).

2.2. Wood modification

2.2.1. Background, description and examples

Wood is one of the most used materials with various applications worldwide. From one hand, the excessive utilization of the virgin sources of the wood and natural forests is a potential danger for environment and should be limited, but on the other hand, there is a huge need for the round timber and other types of lignocellulosic materials to support wood industry. During last century, plantation of the fast growing softwood and hardwood species has highly increased in order to provide the required raw materials needed for industry. Generally, these fast growing species do not have the same quality as good as high quality value tropical timbers. Wood modification is one of the most successful employed technologies for improving some properties of the wood and hence to stay competitive with other mostly non-sustainable materials such as plastics, metals etc. The wood cell walls alter through a chemically, physically or biotechnologically approach after modification treatments (Militz and Hill 2005). Wood modification is an environmental friendly technology aiming to improve the biological durability, decay resistance, dimensional stability, UV resistance and hardness of the wood and it is defined as the action of a chemical, biological or physical agent upon the material, which results in a favorable development of the wood characteristics within the service life of the modified wood. It should be nontoxic and without release of any poisonous substances during service (Kumar 1994, Militz et al. 1997, van Acker and Hill 2003, Hill 2006).

Wood modification can be classified in 4 groups including chemical, thermal, impregnation and

surface modifications (Hill 2006) while several modification methods exist within these

categories with different effects on wood properties (Homan 2004). The principle of different

wood modifications is illustrated in Figure 2.

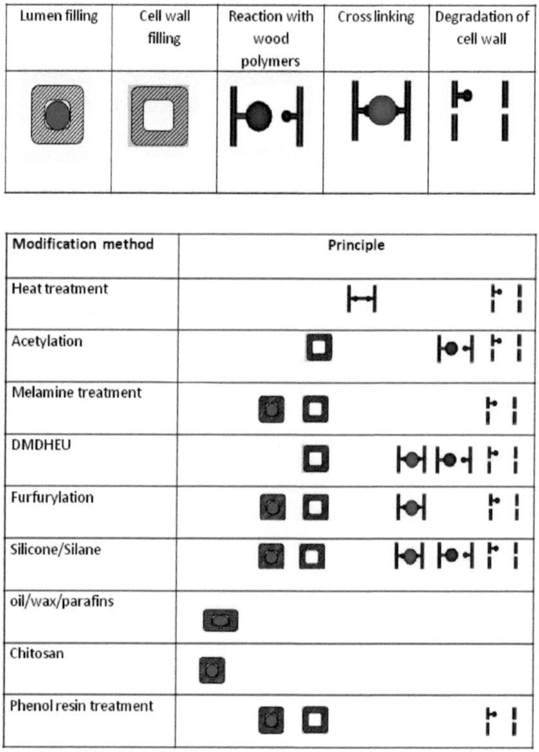

Lumen filling	Cell wall filling	Reaction with wood polymers	Cross linking	Degradation of cell wall

Modification method	Principle
Heat treatment	
Acetylation	
Melamine treatment	
DMDHEU	
Furfurylation	
Silicone/Silane	
oil/wax/parafins	
Chitosan	
Phenol resin treatment	

Figure 2: Wood modification principles (Militz 2014)

2.2.2. Modified species

The modified materials used in different studies within this thesis were solid wood boards of phenol–formaldehyde modified beech (*Fagus sylvatica* L.) in paper 1, furfurylated Scots pine (*Pinus sylvestris* L.), melamine treated Scots pine, and thermally treated Scots pine and beech (*Fagus sylvatica* L.) in paper 2-6 and furfurylated, melamine treated and thermally treated veneers of Silver birch (*Betula pendula* Roth) in paper 7.

2.2.3. Furfurylation

Treatment of the wood with furfuryl alcohol is considered as one of the impregnation modifications. Furfuryl alcohol is obtained from renewable sources, mainly corn cobs. According to the technology of Kebony ASA (Skien, Norway), there are three phases in furfurylation process; impregnation, polymerization and post-curing (figure 3). In the first phases of the treatment, which is a full cell vacuum-pressure impregnation stage, the treatment solution including furfuryl alcohol, citric acids, cyclic carboxylic anhydrides, surfactants and buffering agents (all diluted in water-based solvent systems) is added into a stainless steel reactor to impregnate wood. After impregnation, the polymerization of the penetrated solution molecules inside the wood tissue happens as result of hot steam injection into curing chamber. The last phase is a post curing or final drying step (Westin et al. 2003, Westin et al. 2004, Hill 2006, Treu et al. 2009).

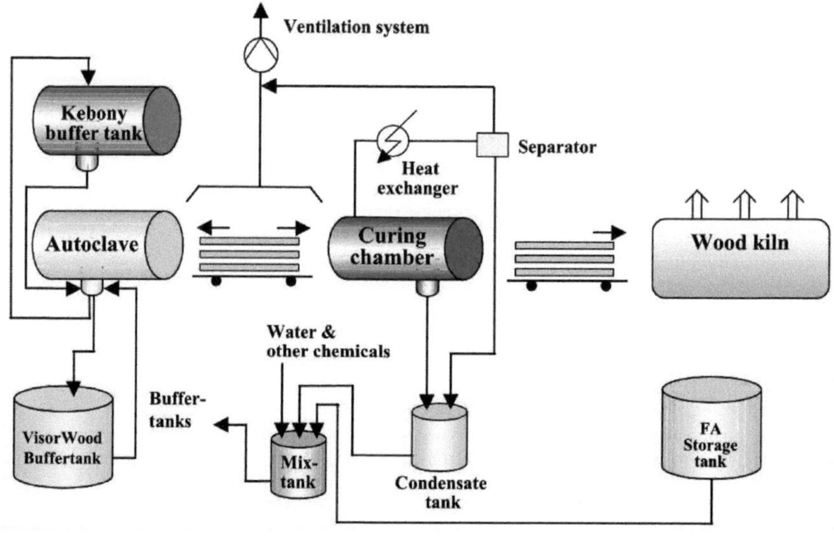

Figure 3: Furfurylation process scheme as suggested by (Westin et al. 2004)

Figure 4 shows the initial polymerization of furfuryl alcohol according to Westin et al. (2003). As it is shown, two competing reactions happen during this process (Figure 4, a and b).Generally, the higher reaction temperatures and concentrations of furfuryl alcohol results in domination of reaction (a) which reduces the formation of ether bridges according to reaction (b). Due to unstable nature of the ether bridge, it undergoes further reaction (Figure 4, c). A three-dimensional polymeric network within the cell wall forms upon further possible cross-linking reactions (Westin et al., 2003, Hill 2006).

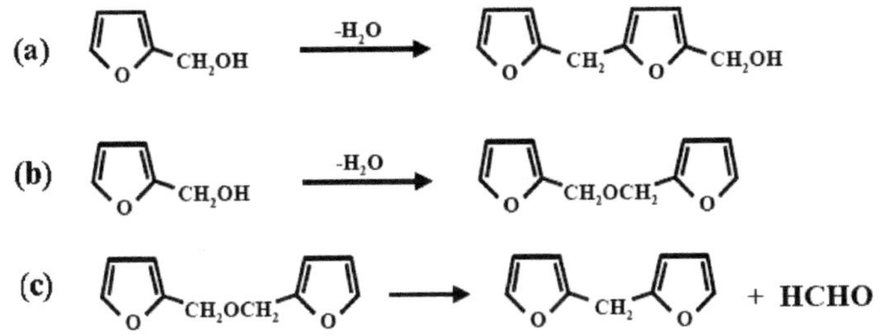

Figure 4: Initial polymerization of furfuryl alcohol (Westin et al. 2003)

The cross-linking reactions which form the three-dimensional polymeric network of furfuryl alcohol within wood cell walls are depicted in figure 5.

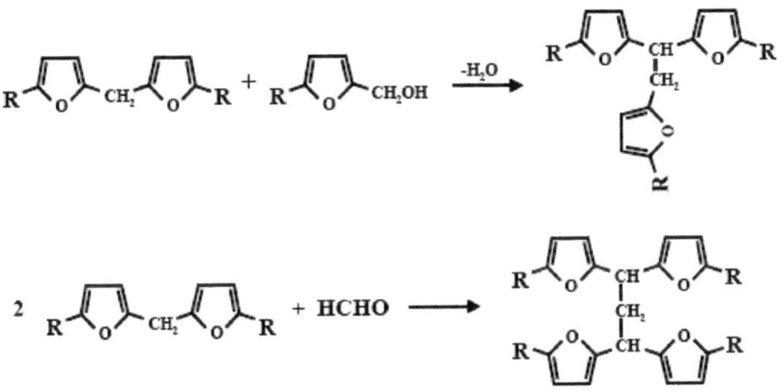

Figure 5: Cross-linking reactions of furfuryl alcohol to form a three-dimensional polymeric network (Westin et al. 2003)

2.2.4. Melamine treatment

N-methylol Melamine (NMM) compounds are used in textile, wood and paper industries as coatings and adhesives (Mark 1965, Hill 2006, Trinh 2009). Modification of the wood with NMM compounds can bring increased dimensional stability (Nicholas and Williams 1987, Militz

1993, Zee et al. 1998, Lukowsky 2002, Nguyen et al. 2007), enhanced resistance against biological degradation such as fungal attack (Videlov 1989, Rapp and Peek 1996, Lukowsky et al. 1999) and improved weathering performance of the wood (Rapp and Peek 1999, Krause and Militz 2004, Hansmann et al. 2005). NMM compounds are able to penetrate, diffuse and polymerize into wood cell walls, resulting in bulking and a cross-linked structure of the cell walls (Gindl et al. 2004, Xie et al. 2005, Trinh 2009, Sint et al. 2013). Although the covalent bonding between wood polymers and NMM compounds has been recorded (Troughton and Chow 1968), there is a general believe that the improved properties of the wood after NMM modification are the result of formation of a three dimensional network of the resin molecules into the cell walls (Lukowsky 1999). The chemical structure of N-methylol melamine is shown in Figure 6.

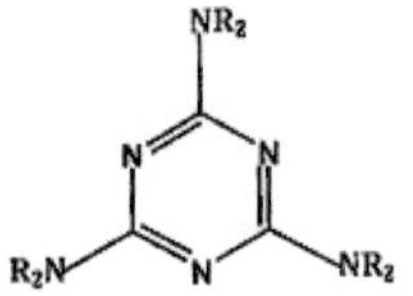

Figure 6: N-methylol melamine (melamine formaldehyde) formula (Mark 1965, Trinh 2009).

Like furfurylation, treatment of the wood with NMM compounds is also classified as an impregnation modification. It is consisted of mixing N-methylol melamine solution with wood in a stainless steel vessel (Figure 7), impregnation using a full cell process with vacuum and pressure and curing in a drying chamber steps (Kielmann et al. 2014, Bastani et al. 2015a).

Figure 7: Impregnation vessel

2.2.5. Thermal modification

Thermal modification is well-known as the nowadays most commercially used wood

modification and it is one of the highly demanded materials for interior and exterior applications

like cladding, decking, flooring and furniture (Militz 2002, Militz and Altgen 2014). The pleasant

coloration of the wood after thermal modification seen by some customers as an advantage of

thermally treated wood comparing to other wood modifications, making it suitable for decorative

applications (Mitsui et al. 2001). However, the mechanical properties of the wood such as shear

strength have shown a considerable reduction after thermal modification as reported in several

studies (Sernek et al. 2008, Sahin Kol et al. 2009, Bastani et al. 2015d). Thermal modification of

the wood normally performs between 160°C and 260°C and alters the hydrophilic characteristics

of the wood as result of thermal degradation of cell wall polymers, mainly hemicelluloses

(Boonstra et al. 2007, Mahnert et al. 2013, Militz and Altgen 2014).

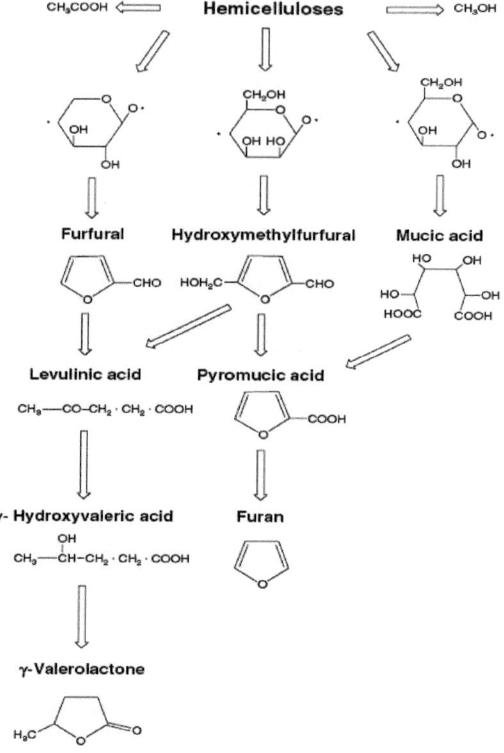

Figure 8: Proposed thermal degradation of hemicelluloses suggested by Fengel and Wegener in 1989 (Hill 2006)

There are various methods of thermal modification on the market, mainly classified based on treatment temperature and applied pressure and presence of the oil and moisture in the process (Militz 2002, Militz and Altgen 2014). The chemical changes and hence the properties of the wood after thermal modification differ between various processes (Tjeerdsma et al. 2002). In this thesis, an industrial scale vacuum press dewatering method (Vacu³) was used for the paper 2-6 and for the paper 7, the thermal treatment was performed using a lab scale heating oven through 9 gradual steps of heating by circulation of hot air and steam in the system.

2.2.6. Phenol treatment

Since the 1930s, impregnation of the wood with low molecular weight, water soluble phenol–formaldehyde (PF) resin has been studied in order to improve dimensional stability of solid wood and also commercial plywood composites, called "Impreg" and "Compreg" (Stamm 1959; Hill 2006; Gabrielli and Kamke 2010). The PF resin penetrates into the porous wood tissue and brings a bulking structure to the cell walls (Stamm and Seborg 1936; Rowell and Banks 1985). Using a low molecular weight PF resin, Ohmae et al. (2002) recorded an anti-shrink efficiency (ASE) of about 74% at 30 % weight percent gain (WPG), as result of bulking and cross-linking of the cell wall. (Ryu et al. 1991) reported an improved fungal resistance of the PF modified wood against white rot and brown rot fungi. It is believed that penetration of PF resin into the cell wall has a major role in efficiency of wood modification (Ryu et al. 1993; Furuno et al. 2004). Modification of the wood with PF resin follows the same principles like other wood impregnation methods including impregnation (vacuum and pressure), curing (gradual drying) and conditioning steps (Bicke and Militz 2015, Biziks et al. 2015).

2.3. Bonding of modified wood and adhesives

2.3.1. Bonding of modified wood

As discussed previously, modification brings several improved properties to the wood as result of chemical and physical changes of its structure. Besides these advantages, modification can alter the bonding behavior of the wood. For example, the less polar and less porous modified wood surfaces may lead to reduced adhesion due to poorer adhesive wetting of the wood and fewer chemical bonds between the two surfaces (Hunt et al. 2007) or the decreased pH and wettability of thermally modified wood might have effect on curing process of adhesives (Boonstra et al. 1998, Sernek et al. 2008, Boruszewski et al. 2011). Bulk treatments (e.g. with resins, waxes) result in reduced water uptake and hence the adhesion of the wood by blocking pathways needed for water (or adhesive) flow and also due to hydrophobic nature of the used materials for modification (Gsöls et al. 2003, Mai and Militz 2004, Weigenand et al. 2007, Zhang et al. 2007, Xiao at al. 2010). As new modification methods are becoming commercially available and by increasing application of modified wood in exterior and interior uses, it is essential to continuously update our knowledge on adhesive bonding of modified wood. During last few years several studies have been carried out on different aspects of bonding of modified wood. Chandler et al. (2005) reported an inferior bonding for acetylated wood bonded with different gluing systems despite of a good degree of adhesive penetration. This finding indicated that lumen penetration does not always correlate with bond strength. The additional OH- groups provided by oxide modifications with butylene and propylene oxide, available for hydrogen bonding with the adhesive, did not lead to a better adhesion than acetylated wood (Brandon et al. 2005). van der Zee et al. (2007) achieved an acceptable gluing performance for furfurylated wood bonded with emulsion polymer isocyanate and melamine urea formaldehyde adhesives, even though the glue line failure increased under wet conditions. Kurt et al. (2008) reported on poor bonding properties of silicone modified wood due to negative effect of silicone compounds on wetting of the wood surface. They suggested using hydroxymethylated resorcinol coupling agent to improve bonding performance of the silicone modified wood. Sernek (2002) did not find any influence of thermal modification on adhesive penetration and also any relevance between bonding performance and penetration but he and his co-workers reported a positive effect of the reduced shrinking and swelling stresses of modified wood on bonding performance (Sernek et al.

2008). It has been demonstrated in several studies that the type of selected adhesive has a major role on bonding performance of modified wood as reported for phenol-formaldehyde treated (Adamopoulos et al 2012), thermally modified (Boonstra et al. 1998, Sernek et al. 2008, Sahin Kol et al. 2009), furfurylated (van der Zee et al. 2007), acetylated (Vick and Rowell 1990, Vick et al. 1993, Frihart et al. 2004) and silicone modified (Kurt et al. 2008) wood. Although, there are several different studies on bonding properties of the modified wood, there is still a lack of comprehensive focused knowledge on different aspects of bonding ability and performance of different types of modified wood. As mentioned earlier, this study examined the bonding properties of the modified wood through three different approaches: investigation of the water uptake and surface properties, mechanical testing and optical observation of adhesive penetration into modified wood structure.

2.3.2. Wood adhesives

Generally, the term adhesive refers to the substance capable of keeping materials together by surface attachment (Pizzi and Mittal 2003, Ebnesajjad 2008). Adhesive performance is the combination of different involved sciences which makes it difficult to understand how an adhesive works. Due to several different applications of adhesives in wood products, there is a wide variety of adhesive types used in wood industry (Vick 1999). One classification divides adhesives into coldset (set at room temperature) and hot set (set by heating) groups while another classification categorize adhesives into two main groups, thermosetting and thermoplastic. Thermosetting adhesives cure via an irreversible chemical reaction while thermoplastics set through losing of solvent or cooling and can soften again by adding solvent or re-heating. Frihart (2005) classified wood adhesives into structural, semi-structural and non-structural groups. The examples of each group are given in Table 3.

Table 3: Classification of wood adhesives according to structural performance at different environmental exposure levels

Structural integrity	Service environment	Adhesive type
Structural	Fully exterior (withstands long-term water soaking and drying)	Phenol-formaldehyde
		Resorcinol formaldehyde
		Phenol-resorcinol-formaldehyde
		Emulsion polymer/isocyanate
		Melamine-formaldehyde
		Melamine-urea-formaldehyde
	Limited exterior (withstands short-term water soaking)	Isocyanate
		Epoxy
		Urea formaldehyde
	Interior (withstands short-term high humidity)	Casein
Semistructural	Limited exterior	Cross-linked polyvinyl acetate
		Polyurethane
Nonstructural	Interior	Polyvinyl acetate
		Animal
		Soybean
		Elastomeric construction
		Elastomeric contact
		Hot-melt
		Starch

(Vick 1999)

2.3.2.1. EPI

The emulsion polymer isocyanate (EPI) glues are generally two- component, dispersion based adhesives which are used in bonding of panels, parquet, window frames, furniture parts, plywood, glulam beams, plastics or metals to the wood surface and OSB web into the flange in production of I-joists. This adhesive consists of water-emulsifiable isocyanate and hydroxyfunctionalized emulsion latex components. The emulsion state of EPI permits to use polymers with higher molecular weight in adhesive composition while it is still possible to maintain the viscosity of the adhesive low, resulting in easier application of the glue (Frihart 2005). EPI is a coldset adhesive with a fast curing speed and light colored glueline with a high flexibility and low creep. The isocyanate component in EPI acts as crosslinking agent, bringing a high water and heat resistance to the resin. EPI adhesives have a complex curing behavior including formation of the emulsion resin along with chemical reactions of the highly reactive isocyanate towards water, hydroxy-, amines- and carboxy-groups (Grøstad and Pedersen 2010).

2.3.2.2. PVAc

Polyvinyl acetate (PVAc) is a linear, rubbery synthetic polymer with an aliphatic backbone which sets by losing water (Murray 1997). This glue is a water-borne and cold set adhesive with $(C_4H_6O_2)n$ formula (Figure 9). This thermoplastic resin was first discovered by Fritz Klatte in 1912 in Germany (Deutsche Reichs Patent no. 281687, 4 July 1913). It is known as white glue or carpenter's glue with a wide range of application in wood assemblies (e.g. furniture, construction) , paper, cloth and building stone (as a consolidation agent). Due to the water-borne nature of PVAc, it flows easily into cell lumens but hardly to the cell walls, owing to its high molecular weight. Having high acetate group contents and a flexible backbone, PVAc forms several hydrogen bonds with various wood fractions (Frihart 2005).

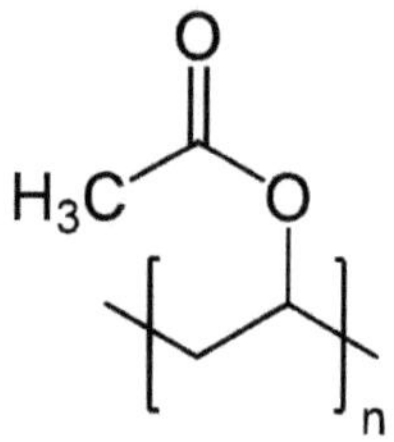

Figure 9: Chemical structure of polyvinyl acetate (PVAc) (Pizzi and Mittal 2003, Ebnesajjad 2008)

2.3.2.3. PU

Polyurethane (PU) is known as a polymer with a chain of organic units linked by carbamate connections. This polymer was first produced in 1937 by Otto Bayer and coworkers (Bayer 1947) and is formed by reaction of isocyanate with polyol (Figure 10).

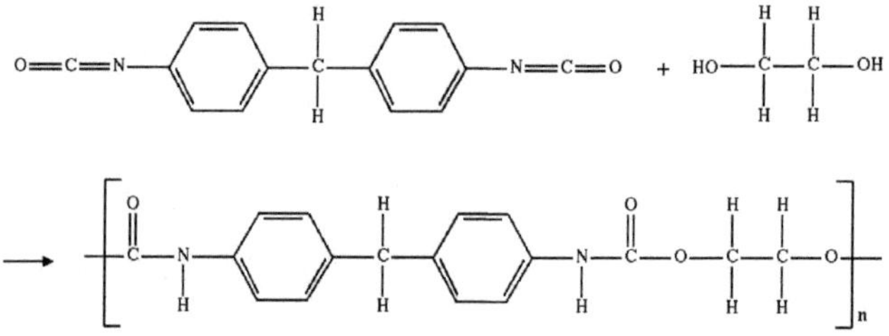

Figure 10: Polyurethane synthesis (Pizzi and Mittal 2003, Ebnesajjad 2008)

Polyurethane polymers are among the reaction polymers (Gum et al. 1992) with a good strength, flexibility, impact resistance and are available in both thermosetting and thermoplastic forms. They are used in textile and packaging markets as well as in production of flexible and rigid foams, insulation panels, surface coatings, wheels and tires and high performance adhesives for bonding glass, ceramics, plastics and wood (Oertel 1985, Woods 1990, Harrington and Hock

1991, Lay and Cranley 2003). PU adhesives are either one- or two-component gluing systems. In one-component system, the isocyanate groups react with moisture of the glued material and further form a crosslinked network (Frihart 2005).

2.3.2.4. PF

Among different wood adhesives, formaldehyde-based glues are the most important adhesives in wood based panels industry. Phenol formaldehyde (PF) resins are considered as the oldest type of synthetic polymers which were developed by beginning of the 20th century. They mainly obtained by the reaction of phenol with formaldehyde. According to production condition and formaldehyde/phenol (F/P) ratio in adhesive composition, PF resins have two basic types of pre-polymer glues, novolaks type which produces under acidic conditions with (F/P) ratio of less than 1 and resole type which produces under basic conditions with F/P ratios of higher than 1. The resole type is more common to use in wood applications due to good wetting properties and its heat activated nature of curing which provides enough time for assembly of adherents. After application of adhesive (pre-polymers) to the wood surface and by heating of the glue in assembly, the polymerization happens which forms a cross-linked polymer network (Detlefsen 2002, Dunky 2004, Frihart 2005). The chemical reactions of PF resins are well described by Detlefsen (2002) and Pizzi (2003). Phenolic resins are used in many products such as phenolic laminates (paper and glass phenolics), phenolic micro-balloons, snooker balls, brake pads, brake shoes and clutch disks. In wood products they can be found in manufacturing of exterior plywood, laminations and wood composites. They have a unique durability and produce a high bonding strength which can fulfil almost all requirements needed by most of the durability testing and wood applications.

2.3.2.5. PRF

Phenol-Resorcinol Formaldehyde (PRF) adhesives belong to resorcinol adhesives group and are widely used in wood lamination, finger jointing and production of exterior glulam and timber structures. They are provided as a liquid compound and a hardener should be added to the adhesive before application. PRF resins cure at room temperature and form very durable bonds with high resistance to failure and degradation. Generally, the biggest disadvantage of resorcinol based adhesives is the cost of resorcinol. In order to lower the cost of the glue, but still to maintain the room temperature curing properties, phenol-resorcinol-formaldehyde (PRF) adhesives were introduced to the market. Theses adhesives have a long assembly time which requires clamping of the adherents (Pizzi 2003, Frihart 2005). The PRF adhesives are mainly produced through grafting of the resorcinol into the active methylol groups of the low-condensation resoles acquired by the reaction of phenol with formaldehyde (Ormstad et al. 2002)

2.4. Aim of Study

Although the application of modified wood (both impregnated and thermally modified) has highly increased during last years, there is still a big challenge to improve and expand the technological aspects of using modified wood such as its bonding process where there is a lack of comprehensive knowledge on bonding properties. Several studies have been performed on bonding properties of the modified wood during last years, but they were mostly restricted by the number of investigated variables, the type of the used modifications and adhesives in each study. This present thesis tried to add knowledge on bondability of different types of modified wood glued with different adhesive systems by considering various aspects: water related characteristics, mechanical performance and optical (fluorescence microscopy and X-ray micro-computed tomography) observation of adhesive penetration into modified wood structure. Scots pine, beech and birch as common wood species used in furniture and construction applications, furfurylation, melamine, phenol and heat treatment being among the commercial wood modifications, and three main coldset solid wood adhesives (two water-borne, one moisture curing) used widely in woodworking industry and a hot curing phenol formaldehyde adhesive were selected for this study. This combination of wood species, modification methods and adhesives could provide sufficient information on bonding properties of the modified wood.

The aim of this study was the development of the knowledge on bonding related properties of modified wood through following steps:

- Evaluation of the effects of phenol formaldehyde modification of the beech wood on its bonding performance.
- Study the water uptake and wetting behaviour of furfurylated, N-methylol melamine modified and thermally modified wood.
- Evaluation of the gross adhesive penetration in furfurylated, N-methylol melamine modified and thermally modified wood.
- Study of adhesive bondlines in modified wood with fluorescence microscopy and X-ray micro-computed tomography.
- Examination of the shear strength of furfurylated, N-methylol melamine and thermally modified wood bonded with three conventional wood adhesives.
- Investigation on the effect of open assembly time and equilibrium moisture content on the penetration of polyurethane adhesive into thermally modified wood.
- Study of the development of bonding strength of modified birch veneers during adhesive curing.

2.5. Evaluation methods used in this research

2.5.1. Evaluation of the wood surface and water related properties

2.5.1.1. Capillary water uptake

Due to the porous character of wood, the water flows through capillary force in wood, and also

diffuses into the cell walls. The capillary water uptake test method provides estimations of its

water uptake behaviour in exterior use as well as of the localisation of chemicals in different

morphological regions of modified woody tissues, which may block the fluid flow path (Rowell

and Banks 1985). Therefore, in order to evaluate the water related properties of the modified

woods and hence to predict their comparable gluing performance due to the liquid nature of

adhesives, the capillary water uptake test was carried out on modified woods according to DIN

52617 (1987), along the longitudinal direction on samples with dimensions of $20 \times 20 \times 200$ mm^3

($R \times T \times L$) and along the tangential and radial directions with samples with dimensions of $40 \times$

40×40 mm3 ($R \times T \times L$). For each case (e.g. treatment and direction) ten replicates were used.

According to the method, the samples were sealed with a water resistant coating along all faces

except two (transverse, radial or tangential) to test each time the longitudinal, tangential and

radial water uptake. The samples were then put in a water vessel in contact with a saturated

sponge to absorb water through the open faces (Figure 11).

Figure 11: Experimental set-up for the capillary water uptake test (longitudinal samples)

The water level was kept 2 mm above the sponge surface, so as a non-sealed surface of the samples was only partially submerged in water to assess the intensity of water absorption. Following DIN 52617 standard, the water uptake was measured after specific time intervals of 1, 2, 3, 4, 8, 12, 24, 48, 72, 96, 120, 168 and 336 h and the water uptake coefficient was calculated for every direction by using the equation below:

$$w_t = \Delta W_t \, t^{-0.5} \qquad (1)$$

where

w_t = water uptake coefficient (kg m^{-2} h$^{-0.5}$)

ΔW_t = difference in mass of water uptake per unit area of sample surface (kg m^{-2}) between start and time t

t = the measuring time (h)

2.5.1.2. Contact angle measurement

As mentioned previously, the wettability of the wood surface is very important in gluing process: the higher wettability of the wood surface the better spread of the adhesive on wood surface and hence better bonding. The wettability can be determined by parameters which indicate to the molecular polar or non-polar interactions between liquids and solids such as contact angles, surface free energy and work of adhesion (Mantanis and Young 1997, de Meijer et al. 2000, Wålinder and Bryne 2006). In this work, the contact angles of water and diiodomethane probe liquids were measured at University of applied sciences and arts (HAWK institute-Göttingen) on the surface of the modified woods to determine their hydrophobic behavior and wettability. The measurement was done according to sessile drop method using a G 10 device (Krüss GmbH, Hamburg, Germany) and DSA 1 software on planed tangential and radial surfaces of wood samples. The dosing volume of probe liquids was 10 µl in each measurement and 5 s after deposition of droplet, the contact angle value was taken. The primary pre-tests showed that this period of time was required to achieve droplet stabilization in both unmodified and modified wood and also to minimize the risk of solvent contamination with wood extractives (Wålinder and Johansson 2001). Figure 12 shows the measuring device composed of dropper needle, recording camera and the adjustable platform where the sample should be placed (in the middle of the picture).

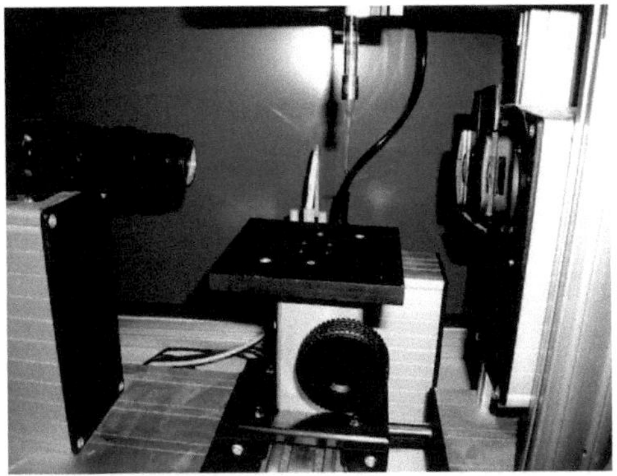

Figure 12: Contact angle measurement device

In order to minimize the effects of structural and chemical variations of the wood samples, for each sample, 20 representative measurements were carried out at different positions.

2.5.1.3. Surface energy of the wood

In addition to the contact angle measurement, the surface properties of the modified wood materials were also determined using surface energy measurement method. Different methods are given in the literature to calculate the surface energy of solids based on contact angle data, mainly using the Young equation as a basis (Żenkiewicz 2007). In this work, the geometric mean approach which was developed by Owens-Wendt (Owens and Wendt 1969) was used to calculate the surface energy of different modified materials. This method is a commonly used procedure for calculation of wood surface energy. The mean values of contact angle measurements were used to calculate the corresponding surface energy by following the Owens–Wendt approach on the basis of Young's equation:

$$\gamma_s = \gamma_{SL} + \gamma_L \cos \theta \qquad (2)$$

where

γ_s = surface energy of the solid (mN m^{-1})

γ_{SL} = interfacial energy between solid and liquid

γ_L = surface tension of the liquid

θ = contact angle

According to this approach, the total surface energy γ^T is consisted of disperse or non-polar component γ^D and a polar component γ^P:

$$\gamma^T = \gamma^D + \gamma^P \qquad (3)$$

At least two measuring liquids (usually water and diiodomethane) are needed to define the polar and non-polar components of surface tension. However, the more liquids that are used the more reliable the results become. As reported by van Oss et al. (1990), the surface tension components γ^D and γ^P for the probe liquids are respectively 21.8 and 51 mN.m^{-1} for water, and 50.8 and 0 mN.m^{-1} for diiodomethane.

2.5.2. Optical Evaluation

2.5.2.1. Light and fluorescence microscopy

There is a common belief that penetration of the adhesive into the wood structure plays an important role in bonding process and production of glued wood-based panels and products by affecting the bond quality (Frihart 2005, Kamke and Lee 2007). A proper quantity of adhesive penetration is always required to produce effective bonding of wood particles and layers by providing the surface contact needed for adhesion (Pizzi 1994). Inadequiet penetration of adhesive brings about minimal surface contact which creates thick film of adhesive on the surface while over penetration leaves starved bondlines, thus leading to poor bonding in both cases (Johnson and Kamke 1992). Due to the importance of the adhesive penetration in wood bonding, in this work, light and fluorescence microscopy techniques were used to observe adhesive penetration into the porous wood structure. Using a Reichert-Jung sliding microtome, 20-40 µm-thick sections exposing a bondline with a cross-sectional surface at various random positions of the glued specimens were cut. For light microscopy the used samples kept unstained but for fluorescence microscopy the sections were stained with proper staining solutions for each type of adhesive and further washed with water and ethanol in several steps. The prepared sections were placed on glass slides and examined via conventional light microscope in the case of unstained samples and under an Eclipse 50i fluorescence microscope with appropriate filter sets, equipped with a Sight DS-5M-L1 digital camera for stained samples. The NIS-Elements F software was used for image analysis (all Nikon, Düsseldorf, Germany). Figure 13 illustrates different components of a fluorescence microscope.

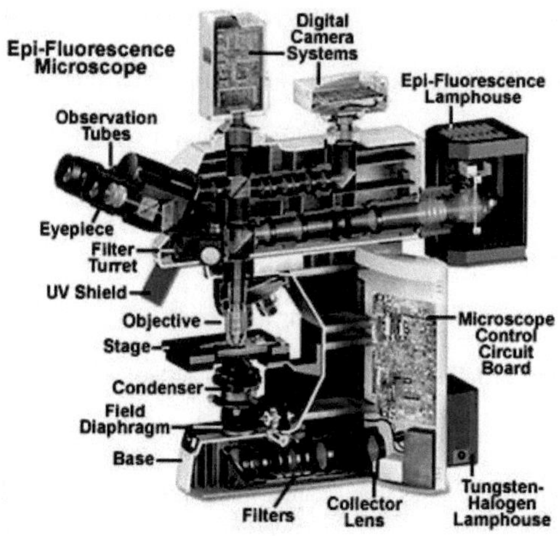

Figure 13: Fluorescence microscope components (MicroscopyU- Nikon Instruments, Inc.)

Penetration of adhesives into wood structure was measured as effective (EP) and maximum penetration (MP) following the method described by Sernek et al. (1999). According to this method, EP is the whole area of adhesive found in the interphase region of the glue line divided by the width of the glue line. Unfilled cell lumens and cell walls are not considered in calculation of EP. MP is the mean distance of penetration of the five most remote adhesive objects detected inside the field of view. The used software has the possibility to distinguish between the bright adhesive objects and the darker background and to calculate the required statistical parameters. Ten regions of interest with dimensions $1100 \times 600 \ \mu m^2$ (W × H) were selected for each case to measure EP and MP using the formulas below:

$$EP = \frac{\sum_i^n Ai}{Xo} \qquad (4)$$

where

EP = effective penetration (µm)

Ai = area of adhesive object I (µm2)

n = number of objects

Xo = width of the maximum rectangle determining the measurement area (1100 µm)

$$MP = \frac{\sum_i^5 (yi+ri+yo)}{5} \qquad (5)$$

where

MP = maximum depth of penetration (µm)

yi = centroid of adhesive object i representing the deepest penetration (µm)

ri = mean radius of adhesive object I (µm)

yo = reference y-coordinate of the glue line interface (µm)

By selecting these regions of interest it is possible to cover a large area in micron level scale and each EP and MP value obtained by this method represented the average of 50 measurements.

2.5.2.2. X-ray micro-computed tomography

The high amount and very deep penetration of PU adhesive into thermally modified wood observed by fluorescence microscopy led to a more detailed three-dimensional (3D) visualisation of penetration of this adhesive by means of X-ray micro-computed tomography (XµCT).

Scanning of the wood sample was done using a sub-micro-focus XµCT system nanotom® s (phoenix|x-ray, GE Sensing & Inspection Technologies GmbH Wunstorf, Germany, see figure 14).

Figure 14: X-ray micro-computed tomography device

The principle of XµCT is based on the attenuation of X-ray radiation throughout the sample. The attenuation coefficient for X-ray is characterized by density and elemental composition of the sample material. The different X-ray absorptions are saved in a grayscale-based two-dimensional (2D) projection. Thereby, grayscale values of each pixel define the average value of attenuation coefficient. In this method, the dark pixels represent regions within the sample with less X-ray absorption (e.g. air) whereas white or light gray pixels represent dense areas (e.g. wood or glue). Using a set of 2D radiographs, a 3D imaging of the sample volume can be reconstructed, and in order to achieve a more detailed observation, several sub-volumes should be created to choose the best one. Finally, by using analysis software Avizo® fire 7.1 (VSG, Mériganc Cedex, France), the 3D reconstruction can be done.

2.5.3. Evaluation of mechanical properties

A very traditional and practical method of the measurement of bonding quality of the glued

assemblies is the mechanical testing of the bonded joints where the performance of the wooden

structure can be evaluated directly based on mechanical modules. In this work, the tensile shear

strength of solid wood assemblies was measured according to the European Standard (1992) EN

302-1 using Zwick/Roell Z010 universal testing machine (Figure 15, left) while the bonding

strength of veneers were measured by using an automated bond evaluation system ABES

(Incorporated, Corvallis, OR, USA, see figure 15, right) following the formula below:

$$Bonding\ strength\ (N/mm^2) = \frac{Force\ (N)}{Bonded\ area\ (mm^2)} \qquad (6)$$

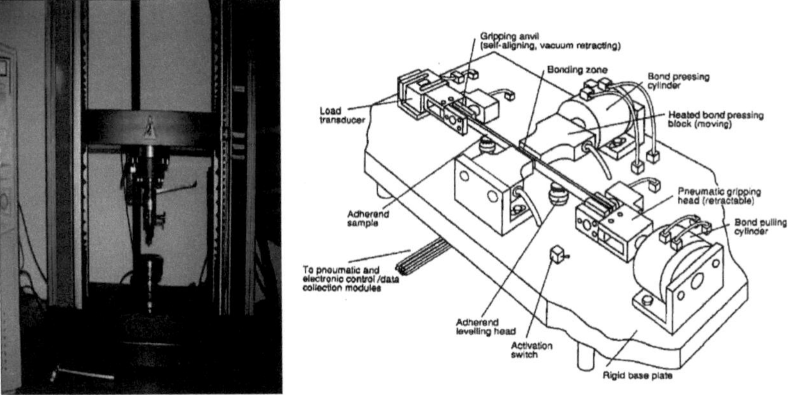

Figure 15: Universal testing machine (Zwick GmbH & Co. KG) in the left and automated bond evaluation system (Humphrey 1993, Dunky 2004) in the right.

3. Results and discussion

3.1. Evaluation of the effects of phenol formaldehyde modification of the beech wood on its bonding performance

When using modified wood for structural applications, it is important to examine its mechanical properties to determine how modification can affect bonding strength of the wood. Modification of the wood with phenol–formaldehyde (PF) resin has a great potential for commercialisation owing to the enhanced properties of the wood after this modification, the relatively low price of PF resin and the rather easy feasibility of the process. Samples of PF treated beech wood (*Fagus sylvatica* L.), in 10 and 25 % concentrations, bonded with phenol resorcinol formaldehyde (PRF) and polyvinyl acetate (PVAc) adhesives were tested according to the European Standard (1992) EN 302-1 to determine their shear strength and wood failure percentage compared to untreated wood. It was assumed that impregnation of the wood with chemicals such as PF solution may have effect on penetration of adhesives into wood structure through adding chemicals on wood tissue pathways. Therefore the effect of PF-modification on penetration of the used adhesives was also investigated by using light and fluorescence microscope. In general under both dry and wet testing conditions and for both treated and untreated woods, samples bonded with PRF adhesive represented a better bonding performance than samples glued with PVAc. The results of dry condition test indicates that PF treated wood bonded with the two adhesive systems could be used for interior applications regardless of wood treated with 25 % PF solution and PVAc adhesive while under wet condition, only samples treated with 25 % PF solution and glued with PRF provided acceptable shear strength. PVAc bonded samples showed a very weak bonding strength and very low wood failure values under wet condition testing. Treatment of beech with PF solution caused a reduction of adhesive penetration into wood structure for both adhesive systems which might have negative effect on bonding performance. This decrease was more obvious at 25 % PF concentration than 10%. Furuno et al. (2004) reported that deposition of PF resin into the cell wall and lumens affected adhesive penetration at 25 % PF concentration where it had a negative effect only on bonding of the samples glued with PVAc. In another study, Treu et al. (2009) reported on the reduction of water uptake of the wood after furfurylation with higher concentration of the furfuryl alcohol. However, in order to have a better understanding about the liquid uptake behavior and surface characteristics of the modified wood, the capillary water uptake, contact angle and surface energy measurement tests were conducted in further experiments.

3.2. Water uptake and wetting behaviour of furfurylated, N-methylol melamine modified and thermally modified wood

The study of the hygroscopic and water related properties (e.g. water uptake) and surface characteristics (e.g. polarity and wettability) of the modified wood is very important to understand the effect of different treatments on practical applications of the modified wood such as its gluing performance. In general, bulk treatments (e.g. with resins, waxes) by blocking pathways needed for the water flow and also owing to the hydrophobic nature of the using chemicals as well as chemical treatments (e.g. acetylation, furfurylation) or thermal modification by changing the chemical properties of wood cell walls as result of decreasing of hydrophilic agents such as hydroxyl groups lead to decreased water uptake and equilibrium moisture content of wood (Militz 2002, Gsöls et al. 2003, Pétrissans et al. 2003, Mai and Militz 2004). Although the purpose of the treatments is attained, resulting in improved dimensional stability and durability of wood, issues with practical aspects of using modified wood such as coating, painting, and gluing might occur by the more hydrophobic wood surfaces. In order to examine the water uptake and wettability of the wood surface after modification, the capillary water uptake (according to DIN 52617), contact angle of water and diiodomethane (using sessile drop method) and surface energy (based on Owens–Wendt approach) were measured for furfurylated Scots pine (*Pinus sylvestris* L.) treated with two concentrations of furfuryl alcohol which are commercially known as FA 40 and FA70 solutions, melamine treated Scots pine impregnated with solutions of 10%, 20% and 30% N-methylol melamine (NMM) solid content, and thermally modified Scots pine and beech (*Fagus sylvatica* L.) treated at 195°C and 210°C. For water uptake experiment, the specimens were sealed with a water resistant coating along all faces except two faces exposed to the saturated sponge to absorb water and the uptake was measured after the time intervals of 1, 2, 3, 4, 8, 12, 24, 48, 72, 96, 120, 168 and 336 h in all three anatomical directions of the wood. A substantial decrease in water uptake was noted after all modifications and along all three directions both after short (24 h) and long exposure times (168, 336 h) which generally indicates to the hydrophobic behavior of the wood after modification. A significant reduction of water uptake (71-89 %) was seen for furfurylated wood which was a bit more obvious in the case of samples treated with higher concentration of furfuryl alcohol. This reduction effect for higher levels of furfurylation was also shown by Treu et al. (2009) for beech wood. This could be due to the blocking of the anatomical pathways of the wood tissue by chemicals or through cell wall bulking effect. For NMM-modified wood, the solution concentration had no effect on the reduced water uptake of the modified wood. The reduction mechanism of water uptake in NMM-modified wood is similar to furfurylated wood (Mahnert et al. 2013; Sint et al. 2013; Kielmann et al. 2014). For thermally treated wood, the highest reduction of water uptake was noted in longitudinal direction of the beech (81–89 %), while radial water uptake was less affected by thermal treatment in both Scots pine and beech species. With the exception of radial uptake of Scots pine, thermal modification at 195°C gave better results than the treatment at 210°C. However, wood defects occurring during thermal treatment (e.g. cracking and broken cell walls) might also have influence on water uptake. For contact angle measurement experiment, 20 different positions of

each sample were examined to determine an average contact angle value on planed tangential and

radial surfaces. This value was taken 5 s after droplet deposition, the required time for

stabilization of droplet on the wood surface. The hydrophobicity of the modified wood surface

was also proved by contact angle data. However, some exceptions were observed, mainly for

thermally modified wood. The load of furfuryl alcohol and NMM had no effect on contact angles,

and so it was for the treating temperatures in the case of thermally modified wood. In general, the

tangential surfaces showed the highest increase of contact angles for both furfurylation and NMM

treatment, while the temperature had no effect on contact angles between the radial and tangential

thermally modified surfaces. Due to the large decrease of the polar components of the measured

surface energy and the small increase of the disperse ones, a non-polar property was observed for

both radial and tangential surfaces after all modifications where no significant differences were

seen in surface polarity between different treatments, intensities, and wood surfaces. The

observed changes in the hygroscopic behaviour and wetting characteristics of the wood caused by

different modifications can influence on the interaction between wood and polymers (e.g. glues,

paints, coatings) and the adhesion performance.

3.3. Gross adhesive penetration in furfurylated, N-methylol melamine modified and thermally
modified *wood*

The study of the penetration of adhesive into wood structure is important because a proper

amount of adhesive penetration is always required to produce effective bonding of wood particles

and layers by providing the necessary surface contact for mechanical interlocking, covalent

bonding or secondary interactions which can develop the bonding strength of the adhesive to the

wood (Pizzi 1994, Frihart 2006). Insufficient adhesive penetration creates inferior surface contact and a thick film of the adhesive on the surface while over penetration leads to the so called starved bondlines, thus leading to poor bonding in both situations (Johnson and Kamke 1992). With increasing use of modified wood for different applications and in order to have a better utilization of this material for bonding process, it is necessary to investigate the interaction between adhesive systems and modified wood tissue. In fact, wood modification creates a new surface and tissue for bonding, e.g. the chemical modification of wood components, particularly hemicelluloses and lignin, happening during thermal modification, reduces the hygroscopicity of the wood and therefore changes the distribution and penetration of the adhesive in wood or impregnation treatments with different chemicals can also affect bonding by providing deposition of the chemicals into the cell walls and lumens and hence changing adhesive penetration pattern into the wood. The same types of modified wood as the previous study on water uptake and wetting behaviour of modified materials were used in this study, emulsion polymer isocyanate (EPI), one-component polyurethane (PU) and poly (vinyl acetate) (PVAc) adhesives were spread uniformly on the tangential surfaces of the wood samples and the glued specimens were left aside for 1 week at room temperature (20°C), thus letting adhesives to penetrate as much as possible. This timing is required since different adhesives may have different penetration behavior over the time. Then the radial penetration of adhesives into wood structure was determined as described in section 2.5.2.1 by using fluorescence microscopy, based on measurements of effective (EP) and maximum penetration (MP). This method provides very helpful information, as it gives knowledge on both; the amount of penetrated adhesives (EP) and the depth of penetration (MP) in a selected region of interest. It should be noted that only a deep penetration of an adhesive in a certain region of interest doesn't imply a good bonding as it may happen in a low quantity and vice versa a high amount of penetrated adhesive doesn't also indicate on a good adhesion as it may happen only within upper layers of the wood cells close to bondline. In this study, in order to establish the basic knowledge on the interaction between modified materials and adhesive systems, as a first step, the adhesive penetration into modified wood tissue was determined without application of pressure as it was assumed that penetration behavior of adhesives might be different with and without application of pressure.

Modification with higher concentration of furfuryl alcohol (FA 70) decreased the EP of EPI adhesive while an improved penetration was noted for PVAc into furfurylated wood. This increase in EP of the water-based and low viscosity PVAc could be explained by the existence of the hygroscopic buffering salt agents which can be founded in furfurylation process (Bryne and Wålinder 2010), as they might increase wetting of this type of adhesive. All adhesives gained a deeper penetration into wood treated with lower concentration of furfuryl alcohol. The result of the water uptake test in the previous study also indicated a bit more reduction in water uptake of samples treated with higher concentration of furfuryl alcohol. NMM treatment caused a reduction of the EP of EPI and PU adhesives into samples treated with 30% solution but similar to furfurylation, PVAc adhesive showed a higher EP after 10% and 20% NMM treatment. Among used adhesives, PU gained a deeper penetration into NMM tread wood despite of its reduced EP after this modification, possibly due to its lower molecular weight. For all adhesives, Scots pine samples treated at 210 °C showed higher EP values than samples treated at lower treatment temperature which can possibly be explained by the more penetration pathways provided at higher treatment temperature than the lower temperature as it is expected to have more breakage of the cell wall structures at elevated treatment temperatures. In the case of PU and PVAc adhesives, beech samples treated at 195 °C represented higher values of EP. However, thermal modification of the wood is associated with many exceptions in results due to the variety in the used treatment methods and the related parameters of each method, wood species and different chemical reactions happening during thermal modification (Boonstra et al. 1998, Boonstra and

Tjeerdsma 2006). Like NMM treatment, for both thermally modified Scots pine and beech samples and at both treatment temperatures, much deeper penetration was noted for PU adhesive into thermally modified wood tissue than two other used adhesives. The tiny cracks which occur during thermal modification can facilitate the penetration and flow of such small size PU adhesive particles into wood structure. Visual observation and analysis of fluorescence microscopy photomicrographs gave more detailed insight on the modality of adhesive penetration by providing information such as the type of the filled cells (e.g. vessel or tracheid and whether they are fully filled or partly filled), the quality of the bondline (e.g. continuous or non-continuous), the number of the penetrated cells by adhesive layer, the presence of voids in interphase and the region where penetration happened (earlywood-latewood) which can totally provide a better image out of adhesive penetration together with effective and maximum penetration data. The combination of the wood species, modification methods and adhesive systems used in this study provided valuable knowledge on the interaction between three common coldset wood adhesives and modified wood which can be helpful in future research on explaining the bonding behaviour of modified wood.

3.4. Study of adhesive bondlines in modified wood with fluorescence microscopy and X-ray micro-computed tomography

As already discussed in the previous work on gross adhesive penetration, it is obvious that among various involved factors in wood bonding process, penetration of adhesive into the porous wood structure is an important issue since it can influence the bond quality and hence the performance of the whole structure. Due to the improved technological properties which modification can bring to the wood, modified wood has become a very popular material for different purposes during last years, especially for building, decking and outdoor furniture. However, in spite of the need for investigation on the practical applications of the modified wood such as it bonding process and the related issues, there are only few studies that have been done on the penetration of adhesives into modified wood (Sernek 2002, Chandler et al. 2005) where penetration of adhesive might be changed as result of physical and chemical changes of the wood tissue after modification and consequently the bonding performance. In this study, the very same adhesives and modified materials were used as the pervious study on gross adhesive penetration where penetration happed without application of the pressure. In this work the glued samples were pressed using a hydraulic press at 1 N mm^{-2} for 4 h at room temperature and then the penetration of the adhesives into samples were determined following the method described in section 2.5.2.1 by using fluorescence microscopy. Therefore, the relationship between adhesive penetration with and without pressure was also investigated. Besides, the high quantity and depth of penetration of PU adhesive into thermally modified wood, measured by fluorescence microscopy method in both studies, revealed the need of a more detailed three-dimensional (3D) visualisation of penetration of this adhesive into the thermally modified wood structure. This examination was performed via X-ray micro-computed tomography (XμCT) method and the thermally treated Scots pine samples (195°C) were scanned and analyzed as described in section 2.5.2.2.

In the case of pressed samples, the effective penetration (EP) of the PU adhesive significantly decreased after furfurlylation and NMM modification. These two modifications have a similar principle based on filling and blocking of the wood cells with chemical compounds which might limit some of the free possible space for penetration of adhesive. In the case of PVAc adhesive, lower EP values were observed after modification with 30% NMM solution and thermal modification at 210°C for both Scots pine and beech. Like the pervious study where no pressure was applied, in this study furfurylated samples treated with lower concentration of furfuryl alcohol also represented deeper penetration for all adhesives than samples treated with higher load and with exception of thermally modified beech (195°C), PU also gained a much deeper penetration into NMM-modified and thermally modified wood among all used adhesives. Comparison of adhesive penetration with and without pressure revealed that with the exception of EP of PU and EPI adhesives into NMM-modified wood and PVAc into thermally modified beech (195°C), application of pressure provided to some extent different results than the situation when no pressure was applied. The (XµCT) tomography method gave an obvious 3D pattern of the PU adhesive flow into thermally modified Scots pine where the pathways that this adhesive used for flow were depicted clearly. The obtained EP and MP data in this study could be useful in better understanding of the effect of different modifications on interaction between adhesives and wood tissue structure and also to correlate these results to mechanical bonding test to examine the effect of adhesive penetration on bonding performance in further studies.

3.5. Shear strength of furfurylated, N-methylol melamine and thermally modified wood bonded with three conventional adhesives

When using modified wood for furniture, building and structural applications, it is important to examine its mechanical properties such as its shear strength to assure reaching to the minimum required values of mechanical modules needed for these applications. The bondline strength has a direct influence on the final performance and durability of the wood-adhesive joints, and hence of the whole wooden structure (Modzel et al. 2011). As already mentioned in the previous discussions, wood undergoes physical and chemical changes after modification, which might potentially affect its bonding properties. For instance, the less polar and less porous modified wood surface can lead to the formation of less free -OH groups for bonding, resulting in establishment of weaker bonds due to the poorer wetting of the wood surface (Hunt et al. 2007). Therefore, there is a need to study the bonding behavior of the modified wood in order to provide suggestions for effective bonding of different modified materials with various adhesives. The bonding strength of the modified wood has been investigated in several studies (Vick and Rowell 1990, Frihart et al. 2004, Brandon et al.2005, Kurt et al. 2008). In present study, the shear strength of furfurylated, N-methylol melamine and thermally modified wood bonded with EPI, PU and PVAc adhesives were investigated. The same wood materials and adhesive systems were used for this study as the previous work on study of the adhesive bondlines in modified wood and the shear strength of the same assemblies were measured following the European Standard (1992) EN 302-1 as described in section 2.5.3. Then, the shear strength data were correlated to the adhesive penetration data (EP) obtained in the mentioned previous study to examine their relationship. It was assumed that penetration of the adhesive might have influence on bonding strength of assemblies. The relationship between shear strength data and bondline thickness values was also investigated.

The shear strength values were significantly lowered after furfurylation and NMM modification as compared to controls, regardless the treatment intensity and for all adhesives used. In the case of PU adhesive, this reduction was higher for furfurylated samples while in the case of PVAc, melamine treated samples gained a higher reduction in shear strength values. The reduction was mainly attributed to the brittle nature of the wood after these two modifications than the failure of the bondline. The shear strength values were also significantly reduced after thermal medication of Scots pine and beech where the wood failure results suggested the bondline as the weakest link as compared to the wood itself. Among used adhesives, PU showed a better bonding performance with thermally modified beech. In general, the amount of reduction in shear strength values were more or less the same after all modifications which recommend to avoid using modified wood for the highly load bearing applications, especially thermally modified Scots pine . No significant correlation was found between effective penetration of adhesives and bondline thickness data and the shear strength values obtained for each type of modified wood. Chandler et al. (2005) reported that for acetylated wood, good lumen penetration did not result in a desirable bonding strength. However, it should be noted that due to the several involved factors in bonding process, such as the variability of wood species in anatomical features and porosity, the wide variety of adhesive application methods and curing processes, and differences in adhesive chemistries, it is difficult to establish direct relationships between adhesive penetration and bonding performance (Johnson and Kamke 1992; Zheng et al. 2004; Gavrilovic-Grmusa et al. 2012). The different chemical and physical changes of the wood tissue after each modification make this situation more complicated. The other influencing factors such as the decreased chemical bonding or interlocking of adhesives, and also the decreased bonding strength of the brittle modified wood tissue might be the potential reasons for the reduced shear strength values after modification of the wood. It could be helpful to investigate the effect of the other influencing parameters on bonding performance beside examination of the adhesive penetration in further studies.

3.6. Effect of open assembly time and equilibrium moisture content on the penetration of polyurethane adhesive into thermally modified wood

As mentioned in previous discussions, among different involved factors in bonding of the wood, penetration of adhesive into the wood structure is considered as an important influencing factor on bond quality and performance of the glued wooden joints (Gruver and Brown 2006, Kamke and Lee 2007, Guan et al. 2013, 2014). Penetration can vary depending on wood, adhesive and process related parameters (Frihart 2005). Among these influencing parameters, moisture content of the wood and open assembly time of the adhesive can change penetration by controlling the viscosity of adhesive (Scheikl 2002). For example, by increasing moisture content of the wood the viscosity of the adhesive decreases, which improves the adhesive penetration and possibly the bonding strength. In order to avoid bonding problems, a proper moisture content range should be considered for the wooden joints. It is important that the moisture content of the bonded assembly be adjusted as close as possible to the equilibrium moisture content (EMC) of the environment where the final product is going to be used (Marra 1992). Generally, the required moisture content of the wood, needed for an optimal penetration and curing of the adhesive, occurs between 4 and 10%. Choosing an appropriate open assembly time is also essential, since a short time leads to insufficient adhesive penetration, while a long time results in drying of the adhesive on the surface, therefore the bonding performance can be negatively affected in both conditions (Sernek et al. 1999). It is also very important to select a correct open assembly time as it can affect the production time which determines the production costs. Thermal modification of the wood brings a pleasant color to the wood for interior and exterior applications plus the enhanced biological durability and dimensional stability of the wood (Militz and Altgen 2014). On the other hand, it alters the hydrophilic characteristics of the wood as result of the thermal degradation of cell wall polymers, especially hemicelluloses. Reduction of the reactive free hydroxyl groups of cell wall polymers by thermal modification might have effect on gluing or painting process (Kamdem et al. 2002, Nguila et al. 2007, Li et al. 2011). The high amount and depth of penetration of polyurethane (PU) adhesive into thermally modified wood detected in the previous studies (adhesive penetration into modified wood with and without application of pressure) conducted to present study where the effect of two important bonding variables, wood moisture content and open assembly time, on penetration of PU adhesive into thermally modified wood was investigated. In this work before gluing, the thermally modified Scots pine samples were conditioned at 55%, 78% and 94% relative humilities to reach different equilibrium moisture content (EMC) of 5.6, 8.6 and 13.2% respectively for samples treated at 195°C and 5.1, 8.2 and 12.5% respectively for samples treated at 210°C. The conditioned samples were glued and pressed under a hydraulic press at 1 N mm^{-2} for 4 h at room temperature after 15 and 30 min open assembly times and the effective (EP) and maximum penetration (MP) of PU adhesive into thermally modified wood was measured according to the method described in section 2.5.2.1. The levels of the wood moisture content and open assembly time were selected in the frame of the requirements for bonding one-component moisture curing polyurethane adhesive as suggested by adhesive manufacturer. It was assumed this combination of the different open assembly times and equilibrium moisture contents of the wood can lead to different penetration values. It should be noted that the PU adhesive used in this study was a pre-polymerized adhesive, which can penetrate into the cell lumens rather than the cell walls and it flows into the wood structure mainly via the large pits of axial tracheids and ray cells.

According to the obtained results, for the Scots pine samples treated at 195°C and after both assembly times of 15 and 30 min, the EMC level of 8.6% was noted as the ideal moisture content for an effective penetration of PU adhesive. For samples treated at 210°C and after short assembly time of 15 min, the EP of PU adhesive significantly decreased by increasing the EMC of the wood at 12.5%, which suggests to avoid the higher EMC level at short open time of 15 min. With the exception of samples treated at 195°C and with 8.6% EMC, extending open assembly time did not significantly change the EP of PU adhesive. Therefore, it is recommended to preferably use the shorter 15 min open assembly time for applying PU adhesive in thermally modified Scots pine as it can result in saving time and hence to reduce production costs. After shorter open assembly time, both samples treated at 195°C and 210°C with moderate level of EMC (8.6 and 8.2%) gained the highest values of EP while samples with the higher level of EMC (13.2 and 12.5%) achieved the lowest values of EP. The adhesive bondlines observed for the samples with higher EMC levels were thicker as compared to the 30-40 µm bondlines detected for all other samples. Beside examination of the bonding properties of the modified woods bonded with coldset adhesives it could be also useful to investigate the bonding behavior of the modified materials using hot curing adhesives, because such information could provide a better understanding over the entire knowledge on boding ability of modified woods with different adhesives systems.

3.7. Development of bonding strength of modified birch veneers during adhesive curing

Wood furniture and construction industries are established based on cutting wood into smaller pieces and bonding of these pieces together using different adhesives systems. In this way, it is possible to diminish the natural anisotropy of the wood while the optimal dimensions and quality of the final product can be also provided (Hass et al. 2012). In adhesion process, the influencing factors on bond quality have an interaction on each other which forms the final production cost, e.g. a suitable and well adjusted pressing or assembly time leads to reduction of production time which consequently reduces the production costs (Kariž et al. 2009). Thus, improving the bonding influencing factors is very important so that the production process and expenses could be optimized. As modified wood becomes a more demanded material for different applications, there is a need to study various practical aspects of using modified wood such as its bonding process where the challenge is to bond different modified materials as their physical and chemical characteristics are considerably changed by different modifications as compared to the unmodified wood (Boonstra and Tjeerdsma 2006, Rowell 2006, Nguila Inari et al. 2007). In previous studies and discussions in this thesis, the bonding properties of different types of modified wood glued with common coldset wood adhesives were investigated while in present study the bonding strength of different types of modified birch veneers was examined using hot curing phenol formaldehyde (PF) adhesive so that a comprehensive knowledge could be provided on bonding of the modified wood with both coldset and hot curing adhesives. In this study, the development of bonding strength of furfurylated, N-methylol melamine (NMM) and thermally

modified (180 and 220°C) birch veneers glued with hot curing PF adhesive was investigated in

different pressing (20 s , 160s) and open assembly times (20s , 10 min) by using automated

bonding evaluation system (ABES) device as described in section 2.5.3. It was assumed that each

type of modified wood might have different bonding behavior with PF adhesive over different

assembly and pressing times.

According to the obtained results, with extending of the pressing time from 20 s to 160 s, the

bonding strength significantly developed for both unmodified and modified samples after both

open assembly times. The only exception was noted for thermally modified samples (220°C) at

20 s open assembly time. This fact illustrates the importance of the length of pressing for bonding

process of both modified and unmodified wooden materials with such a hot curing adhesive like

PF. The dependency between curing of PF adhesive and pressing time has been already reported

for bonding of the beech samples at 160°C (Jost and Sernek 2009). With the exception of NMM

modified veneers, prolongation of assembly time in the case of 20 s of pressing, did not change

the bonding strength for both modified and unmodified samples while at 160 s pressing, the

bonding strength increased for controls, NMM modified and thermally modified (180°C)

veneers by extending assembly time. The combination of 10 min assembly time and 160 s

pressing time found as the best bonding condition for controls, NMM modified and thermally

modified (180°C) veneers while for furfurylated samples the highest bonding strength was noted

after 20 s of assembly and 160 s of pressing times. In general, for the most of combinations of

open assembly and pressing times, modification affected negatively on the bonding performance

of the veneers, especially in the case of furfurylated and NMM modified samples. In comparison

to the other wood types used in this study, the bonding process of these two types of modified

veneers with PF adhesive is more dependent on pressing time, probably due to poor initial wetting of the surface of the veneers after these modifications. Extending pressing time can highly improve the bonding properties of the furfurylated and NMM modified veneers as result of changing in the chemical combination of the surface of these modified materials after 160 s of pressing. The decreased wettabilty of wood surface after modification was confirmed by contact angle and surface energy measurements on the same modified materials in the second study of this thesis on wetting behaviour of modified materials. The obtained results in this study gave good information on bonding behavior and the interaction between hot curing PF adhesive and different modified wooden materials and it can be used together with the results of previous studies in this thesis, on bonding properties of coldset wood adhesives to provide a vast knowledge on bondability of modified wood.

4. Conclusions and further studies

The bondability of different types of modified wood glued with various cold set and hot curing adhesives was investigated through different approaches in separated studies. According to the obtained results in paper 1, the PRF adhesive showed a better bonding performance with PF treated wood than PVAc. In general, this type of modified wood can be used with both adhesives for interior applications but not for exterior uses, especially PVAc. PF treatment of the beech had a negative effect on penetration of the both adhesives into wood structure.

The results of the paper 2 on water uptake, contact angle and surface energy measurements proved the hydrophobic behavior, reduced wetting and diminished polar properties of the wood surface after modification which can negatively affect its bonding process.

The effective and maximum penetration data obtained in both studies (paper 3 and 4, with and without application of pressure) are valuable information which provide a good knowledge on the effect of different modifications on wood tissue and the interaction between modified woods and adhesive systems, but it should be noted that these information cannot be directly correlated to bonding performance as penetration is only one influencing factor among many involved factors in bonding process. However, the negative effect of modification on penetration of adhesives was seen in several cases. The (XμCT) tomography method proved as a useful tool to provide detailed information on adhesive penetration into wood structure which can be a promising method for future studies on different types of adhesive systems and modified woods provided that the wood and adhesives to be used have different density as this is a prerequisite of this method. According to shear strength results obtained in paper 5, the bonding strength of all glued samples was negatively affected by modification and the worst bonding strength values were noted for the thermally modified Scots pine where the bondline was found as the weakest link as compared to the wood itself. In the case of furfurylation and NMM modification, PVAc showed a better bonding performance than two other used adhesives. The shear strength values showed independent of the effective penetration and bondline thickness data.

According to the paper 6 results, the EMC level of 8.6% was identified as the ideal moisture content for an effective penetration of PU adhesive into thermally modified Scots pine (195°C) while with exception of one case, prolongation of open assembly time form 15 min to 30 min had no effect on EP of PU adhesive into thermally modified Scots pine, which suggests using the shorter open assembly time for applying PU adhesive in thermally modified Scots pine as it can save time and production costs.

The study of boding properties of the modified birch veneers in paper 7 showed that the length of pressing time is very important in bonding of all types of veneers with PF adhesive. Extending assembly time had a positive effect on bonding strength of the most of samples which were pressed for 160 s while it had no effect on bonding of the samples that were pressed for the shorter time of 20 s. Generally, for the most of combinations of open assembly and pressing times, modification had a negative effect on bonding strength of the veneers, especially for furfurylated and NMM modified samples which are more dependent on pressing time than the other types of veneers.

The overall results obtained in this thesis indicate that modified wood has inferior bonding ability than unmodified wood. The decreased wetting and water related properties of the modified wood and the reduced strength of the wood after modification have influence respectively on the bonding ability and the strength of the bonded wood. However, the development of the bonding ability of modified wood is only in its starting point. The future research topic on bondability of modified wood could be the examination of the improving methods of bonding such as application of plasma treatment, using coupling agents, bio products and formulation of the new adhesive systems which can be developed especially for modified woods. On the other hand, it should be also noted that the long term performance of the glued wood products could be much better for modified wood than unmodified wood, due to the high dimensional stability and low water uptake of modified wood, even with less adhesive penetration and less strength of the glue line. Therefore, the examination of the long term performance of the modified wood could be an interesting research topic in future studies.

5. References

Adamopoulos S, Bastani A, Gascon-Garrido P, Militz H, Mai C (2012) Adhesive bonding of beech wood modified with a phenol formaldehyde compound. Eur J Wood Prod 70:897–901

Bastani A, Adamopoulos S, Militz H (2015a) Water uptake and wetting behaviour of furfurylated, N-methylol melamine modified and heat-treated wood. Eur J Wood Prod. 73(5):627-634

Bastani A, Adamopoulos S, Militz H (2015b) Gross adhesive penetration in furfurylated, N-methylol melamine modified and heat-treated wood examined by fluorescence microscopy. Eur. J. Wood Prod. 73:635–642

Bastani A, Adamopoulos S, Koddenberg T, Militz H (2015c) Study of adhesive bondlines in modified wood with fluorescence microscopy and X-ray micro-computed tomography. International Journal of Adhesion Adhesives 68(2016)351–358

Bastani A, Adamopoulos S, Militz H (2015d) Shear strength of furfurylated, N-methylol melamine and thermally modified wood bonded with three conventional adhesives. Wood Material Science and Engineering, DOI:10.1080/17480272.2016.1164754

Bayer O (1947) Das di-isocyanat-polyadditionsverfahren (polyurethane), Angewandte Chemie, 59, 257-272

Bicke S, Militz H (2015) Weathering Stability of PF-treated Veneer Products from Beech Wood. In: Hughes M, Rautkari L, Uimonen T, Militz H, Junge B (Hg.) The Eighth European Conference on Wood Modification. Book of Abstracts. The Eighth European Conference on Wood Modification , S. 101

Biziks V, Bicke S, Militz H (2015) Penetration of phenol formaldehyde (PF) resin into beech wood studied by light microscopy. The International Research Group on Wood Protection, Vina del Mar, Chile; 05/2015

Bolton AJ, Dinwoodie JM, Davies DA (1988) The validity of the use of Sem/Edax as a tool for the detection of UF resin penetration into wood-cell walls in particleboard. Wood Sci Technol 22:345-356

Boonstra MJ, Tjeerdsma BF, Groeneveld HAC (1998) Thermal modification of non-durable wood species. Part 1, The Plato technology: thermal modification of wood. International Research Group on Wood Preservation, Document no. IRG/WP 98-40123

Boonstra MJ, Tjeerdsma BF (2006) Chemical analysis of heat treated softwoods. Holz Roh- Werkst 64:204–211

Boonstra MJ, Van Acker J, Kegel E, Stevens M (2007) Optimisation of a two-stage heat treatment process: durability aspects. Wood Sci Tech 41:31-57

Boruszewski PJ, Borysiuk P, Mamiński MŁ, Grześkiewicz M (2011) Gluability of thermally modified beech (Fagus silvatica L.) and birch (Betula pubescens Ehrh.). Wood Mater Sci Eng 6(4):185-189

Brady DE, Kamke FA (1988) Effect of hot-pressing parameters on resin penetration. Forest Prod. J. 38(11/12):63-68

Brandon R, Ibach RE, Frihart CR (2005) Effects of chemically modified wood on bond durability. In: Frihart CR (ed) Wood adhesives 2005. Forest Products Research Society, San Diego, pp 111–114

Bryne LE, Wålinder MEP (2010) Ageing of modified wood. Part 1: Wetting properties of acetylated, furfurylated, and thermally modified wood. Holzforschung 64(3):295-304

Chandler JG, Brandon LR, Frihart CR (2005) Examination of adhesive penetration in modified wood using fluorescence microscopy. In: ASC spring 2005 convention and exposition, April 17-20, Columbus, OH, 10 p de Meijer M, Haemers S, Cobben W, Militz H (2000) Surface energy determinations of wood: comparison of methods and wood species. Langmuir 16(24):935-939

Detlefsen WD (2002) Phenolic resins: Some chemistry, technology and history. In: Chadhury, M, and Pocius, A.V. (Eds.), Adhesive Science and Engineering – 2: Surfaces, Chemistry and Applications. Elsevier, Amsterdam, chap. 20

Deutsche Reichs Patent no. 281687 (4 July 1913), an abstract of which appears in the Journal of the Society of Chemical Industry (London), vol. 34, page 623 (1915)

DIN 52617 (1987) Bestimmung des Wasseraufnahmekoeffizientenvon Baustoffen. Berlin, Germany

Donath S, Militz H, Mai C (2006) Creating water-repellent effects on wood by treatment with silanes. Holzforschung 60:40-46

Dunky M (2004) Adhesives based on formaldehyde condensation resins. Macromolecular Symposia, 217, 417-429

Ebnesajjad S (2008) Adhesives technology handbook, 2nd ed. William Andrew Inc, Norwich, NY

Esteves BM, Pereira HM (2009) Wood modification by heat treatment: A review. BioResources 41:370-404

European Standard (1992) EN 302-1: adhesives for load bearing timber structures—test methods—Part 1: determination of bond strength in longitudinal tensile shear

Fengel D, Wegener G (1989) Wood: Chemistry, Ultrastructure, Reactions. Walter De Gruyter, Berlin, Germany

Frihart CR, Brandon R, Ibach RE (2004) Selectivity of bonding for modified wood. P Adhes Soc 27(1):329–331

Frihart CR (2005) Wood adhesion and adhesives. Handbook of wood chemistry and wood composites. Boca Raton, Fla.: CRC Press, 2005: pages 215-278

Frihart CR (2006) Wood structure and adhesive bond strength. In: Characterization of the Cellulosic Cell Wall, D. D. Stokke and L. H. Groom (eds). Blackwell Publishing, Ames, IA, pp 241-253

Furuno T, Imamura Y, Kajita H (2004) The modification of wood by treatment with low molecular weight phenol-formaldehyde resin: a properties enhancement with neutralized phenolic-resin and resin penetration into wood cell walls. Wood Sci Technol 37:349–361

Gabrielli CP, Kamke FA (2010) Phenol-formaldehyde impregnation of densified wood for improved dimensional stability. Wood Sci Technol 44:95–104

Gavrilovic-Grmusa I, Dunky M, Miljkovic J, Djiporovic-Momcilovic M (2012) Influence of the viscosity of UF resins on the radial and tangential penetration into poplar wood and on the shear strength of adhesive joints. Holzforschung 66 (7):849-856

Ghosh S, Militz H, Mai C (2009) The efficacy of commercial silicones against blue stain and mould fungi in wood. Eur J Wood Prod 67:159–167

Ghosh S, Militz H, Mai C (2013) Modification of Pinus sylvestris L. wood with quat- and amino-silicones of different chain lengths. Holzforschung 67(4):421–427

Gindl W, Dessipri E, Wimmer R (2002) Using UV-microscopy to study diffusion of melamine-urea-formaldehyde in cell walls of spruce wood. Holzforschung 56:103-107

Gindl W, Hansmann C, Gierlinger N, Schwanninger M, Hinterstoisser B, Jeronimidis G (2004) Using a water-soluble melamineformaldehyde resin to improve the hardness of Norway spruce wood. J Appl Polym Sci 4(93):900–1907

Grøstad K, Pedersen A (2010) Emulsion Polymer Isocyanates as Wood Adhesive: A Review. J. Adhes. Sci. Technol. 24(8-10):1357-1381

Gruver TM, Brown NR (2006) Penetration and performance of isocyanate wood binders on selected wood species. Biores 1(2): 233-247

Gsöls I, Raetzsch M, Ladner C (2003) Interactions between wood and melamine resins—effect on dimensional stability properties and fungal attack. In: Van Acker J, Hill C (eds) Proceedings first european conference on wood modification. Ghent, Belgium, pp 221–225

Guan M, Yong C, Wang L (2013) Shear strain and microscopic characterization of a bamboo bonding interface with poly (vinyl alcohol) modified phenol–formaldehyde resin. J Appl Polym Sci 130(2):1345–1350

Guan M, Yong C, Wang L (2014) Microscopic characterization of modified phenol-formaldehyde resin penetration of bamboo surfaces and its effect on some properties of two-ply bamboo bonding interface. BioResources 9(2):1953-1963

Gum W, Riese W, Ulrich H (1992) Reaction Polymers. New York: Oxford University Press. ISBN 0-19-520933-8

Hansmann C, Deka M, Wimmer R, Gindl W (2006) Artificial weathering of wood surfaces modified by melamine formaldehyde resins. Holz als Roh- und Werkstoff, 64: 198–203

Harrington R, Hock K (1991) Flexible Polyurethane Foams. Midland: The Dow Chemical Company

Hass P, Wittel FK, Mendoza M, Herrmann HJ, Niemz P (2012) Adhesive penetration in beech wood: experiments. Wood Sci Technol 46(1):243-256

Hill CAS (2006) Wood modification: Chemical, thermal and other processes. John Wiley & Sons, Chichester, UK

Homan WJ (2004) Wood Modification, state of the art 2004. Proceedings of the COST E 18 final European seminar on high performance wood coatings exterior and Interior Performance. Paris, France, 26-27 April 2004

Humphrey PE (1993) A device to test adhesive bonds. US patent number 5,176,028. US Patent Office, Washington, DC. USA

Hunt CG, Brandon R, Ibach RE, Frihart CR (2007) What does bonding to modified wood tell us about adhesion? In: Proceedings 5th COST E34 international workshop: bonding of modified wood. Bled, Slovenia, pp 47-56

Johnson SE, Kamke FA (1992) Quantitative analysis of gross adhesive penetration in wood using fluorescence microscopy. J Adhes 40:47-61

Johansson J, Kifetew G (2010) CT-scanning and modelling of the capillary water uptake in aspen, oak and pine. Eur. J. Wood Prod. 68(1):77-85

Johnson SE, Kamke FA (1992) Quantitative analysis of gross adhesive penetration in wood using fluorescence microscopy. J Adhes 40:47-61

Kamdem DP, Pizzi A, Jermannaud A (2002) Durability of heat-treated wood. Holz Roh Werkst 60:1–6

Kamke FA, Lee JN (2007) Adhesive penetration of wood-a review. Wood Fiber Sci 39(2):205-220

Kamke FA, Nairn J. A, Muszynski L, Paris J. L, Schwarzkopf M, Xiao X (2014) Methodology for micromechanical analysis of wood adhesive bonds using X-ray computed tomography and numerical modelling. Wood Fiber Sci 46(1):15-28

Kariž M, Jošt M, Sernek M (2009) Curing of phenol-formaldehyde adhesive in boards of different thicknesses. Wood Res 54 (2): 41-48

Keimel FA (2003) Historical development of adhesives and adhesive bonding. In: Pizzi, A. and Mittal, K.L. (Eds.),

Kielmann BC, Adamopoulos S, Militz H, Koch G, Mai C (2014) Modification of three hardwoods with an N methylol melamine compound and a metal-complex dye. Wood Sci Technol 48:123–136

Konnerth J, Harper D, Lee SH, Rails TG, Gindl W (2008) Adhesive penetration of wood cell walls investigated by scanning thermal microscopy (SThM). Holzforschung 62:91-98

Krause A, Militz H (2004) Cyclic N-Methylol compounds for wood modification. Proceedings of COST Action E29: Innovative Timber and Composite Elements for Buildings Florence, Italy

Kumar S (1994) Chemical modification of wood. Wood Fiber Sci 26(2):270-280

Kurt R, Mai C, Krause A, Militz H (2008) Hydroxymethylated resorcinol (HMR) priming agent for improved bondability of silicone modified wood glued with a polyvinyl acetate adhesive. Holz Roh Werkst 66:305–307

Lay DG, Cranley P (2003) Polyurethane adhesives. In: Pizzi A and Mittal KL (Eds.), Handbook of Adhesive Technology (2nd ed.). Marcel Dekker, New York, chap. 34

Li X, Cai Z, Mou Q, Wu Y, Liu Yuan (2011) Effects of heat treatment on some physical properties of Douglas fir (Pseudotsuga menziesii) wood. Advanced Materials Research. 197-198: 90-95

Lukowsky D (1999) Holzschutz mit Melaminharzen, PhD thesis, University of Hamburg, Germany

Lukowsky D, Büschelberger F, Schmidt O (1999) In situ testing the influence of melamine resins on the enzymatic activity of basidiomycetes. The International Research Group on Wood Protection, IRG/WP 99-30194, Stockholm, Sweden

Lukowsky D (2002) Influence of the formaldehyde content of water based melamine formaldehyde resins on physical properties of Scots pine impregnated therewith. Holz Roh- Werkst 60:349–355

Mahnert KC, Adamopoulos S, Koch G, Militz H (2013) Topochemistry of heat-treated and N-methylol melamine-modified wood of koto (*Pterygota macrocarpa* K. Schum.) and limba (*Terminalia superba* Engl. et. Diels). Holzforschung 67:137-146

Mai C, Militz H (2004) Modification of wood with silicon compounds. Treatment systems based on organic silicon compounds— a review. Wood Sci Technol 37(6):453–461

Mantanis GI, Young RA (1997) Wetting of wood. Wood Sci Technol 31:339-353

Mark R (1965) In: Encyclopedia of Polymer Science and Technolog, Vol. 2, Interscience publishers, New York

Marra A (1992) Technology of wood bonding principles in practice. Van Nostrand Reinhold, New York Handbook of Adhesive Technology (2nd ed.). Marcel Dekker, New York, chap. 1, pp. 1–12

Mayes D, Oksanen O (2003) ThermoWood Handbook. Finnish Thermowood Association, Helsinki, Finland

Militz H (1993) Treatment of Timber with water soluble dimethylol resins to improve their dimensional stability and durability. Wood Science and Technology 27: 347-355

Militz H, Beckers EPJ, Homan WJ (1997) Modification of solid wood: research and practical potential. The International Research Group on Wood Preservation, Document Nr. IRG/WP 97-40098, Whistler, Canada

Militz H (2002) Thermal treatment of wood: European processes and their background. Document IRG/WP 02-40241 33rd Annual Meeting Cardiff

Militz H, Hill C (2005) Modification: Processes, Properties and Commercialisation. In: Militz H, Hill C (Hg.) Wood Modification: Processes, Properties and Commercialisation. Wood Modification: Processes, Properties and Commercialisation, Göttingen, Germany, pp 1-403

Militz (2014) Modified Wood - processes, products and markets.

http://www.unece.org/fileadmin/DAM/timber/docs/tc-sessions/tc-65/md/presentations/17Militz.pdf

Militz H, Altgen M (2014) Processes and properties of thermally modified wood manufactured in Europe. Deterioration and Protection of Sustainable Biomaterials, ACS Symposium Series 1158:269-285

Mitsui K, Takada H, Sugiyama M, Hasegawa R (2001) Changes in the properties of light-irradiated wood with heat treatment: Part 1 Effect of treatment conditions on the change in color. Holzforchung 55:601-605

Modzel G, Kamke FA, De Carlo F (2011) Comparative analysis of a wood: adhesive bondline. Wood Sci Technol 45(1):147-158

Murray GT (1997) Handbook of materials selection for engineering applications. Marcel Dekker, New York

Nguila Inari G, Petrissans M, Gerardin P (2007) Chemical reactivity of heat-treated wood. Wood Sci Technol 41 (2):157–168

Nguyen HM, Militz H, Mai C (2007) Protection of wood for above ground application through modification with a fatty acid modi- fied N-methylol/paraffin formulation. The International Research Group on Wood Protection, IRG/WP 07-40378, Stockholm, Sweden

Nicholas DD, Williams AD (1987) Dimensional stabilization of wood with dimethylol compounds. International Research Group on Wood Protection, IRG/WP/3412

Niemz P, Mannes D, Lehmann E, Vontobel P, Haase S (2004) Untersuchungen zur Verteilung des Klebstoffes im Bereich der Leimfuge mittels Neutronenradiographie und Mikroskopie. Holz Roh Werkst 62(6):424-432

Nussbaum RM (1999) Natural surface inactivation of Scots pine and Norway spruce evaluated by contact angle measurements. Holz Roh Werkst 57:419–424

Oertel G (1985) Polyurethane Handbook. New York: Macmillen Publishing Co, Inc. ISBN 0-02-948920-2

Ohmae K, Minato K, Nonmoto M (2002) The analysis of dimensional changes due to chemical treatments and water soaking for hinoki (Chamaecyparis obtusa) wood. Holzforschung 56:98–102

Ormstad EB, Scheikl M, Pizzi A (2002) Wood adhesion and glued products. In: Dunky M, Pizzi A, Van Leemput M (eds.), State of the Art-Report, COST-Action E13, Part I (Working Group 1, Adhesives), European Commission, Brussels, Belgium

Owens DK, Wendt RC (1969) Estimation of the surface free energy of polymers, J Appl Polym Sci 13:1741-1747

Pétrissans M, Gérardin P, El bakali I, Serraj M (2003) Wettability of heat-treated wood. Holzforschung 57(3):301-307

Paris J. L, Kamke FA, Mbachu R, Kraushaar Gibson S (2013) Phenol formaldehyde adhesives formulated for advanced X-ray imaging in wood-composite bondlines. J Materials Sci 49(2):580-591

Pizzi A (1994) Advanced wood adhesives technology. Marcel Dekker Inc, New York, NY, 289 p

Pizzi A (2003) Resorcinol adhesives. In: Pizzi A and Mittal KL (Eds.), Handbook of Adhesive Technology (2nd ed.). Marcel Dekker, New York, chap. 29

Pizzi, A, Mittal KL (2003) Handbook of Adhesive Technology, 2nd ed. Marcel Dekker, New York

Pries M, Wagner R, Kaesler K-H, Militz H, Mai C (2013) Acetylation of wood in combination with polysiloxanes to improve water-related and mechanical properties of wood. Wood Sci Technol 47:685-699

Rapp AO, Peek RD (1996) Melamine resins as preservatives—results of biological testing. The International Research Group on Wood Protection, IRG/WP 96-40061, Stockholm, Sweden

Rapp AO, Bestgen H, Adam W, Peck RD (1999) Electron energy loss spectroscopy (EELS) for quantification of cell-wall penetration of a melamine resin. Holzforschung 53:111-117

Rapp AO, Peek RD (1999) Melaminharzimprägniertes sowie mit Wetterschutzlasur oberflächenbehandeltes und unbehandeltes Vollholz während zweijähriger Freilandbewitterung. Holz RohWerkst 57(5):331–339

River BH, Vick CB, Gillespie RH (1991) Wood as an adherend. Treatise on adhesion and adhesives.Volume 7. New York : Marcel Dekker, Inc., 1991: 230 pages

River BH (1994) Fracture of adhesively-bonded wood joints. In: Pizzi, A. and Mittal, K.L. (Eds.), Handbook of Adhesive Technology. Marcel Dekker, New York, chap. 9

Rowell RM, Banks WB (1985) Water repellency and dimensional stability of wood. USDA General Technical Report FPL-50:1–24

Rowell R (2006). Chemical modification of wood: A short review. Wood Mater Sci. Eng. 1(1):29-33

Ryu JY, Takahashi M, Imamura Y, Sato T (1991) Biological resistance of phenol-resin treated wood. Mokuzai Gakkaishi 37(9):852–858

Ryu JY, Imamura Y, Takahashi M, Kajita H (1993) Effects of molecular weight and some other properties of resins on the biological resistance of phenolic resin treated wood. Mokuzai Gakkaishi 39(4):486–492

Sahin Kol H, Özbay G, Altun S (2009) Shear strength of heat-treated tali (Erythrophleum ivorense) and iroko (Chlorophora excelsa) woods, bonded with various adhesives. Biores 4(4):1545–1554

Saiki H (1984) The effect of the penetration of adhesives into cell walls on the failure of wood bonding. Mokuzai Gakkaishi 30(1):88-92

Salehuddin A (1970) A Unifying Physico-Chemical Theory for Cellulose and Wood and Its Application in Gluing. Thesis submitted to North Carolina State University, Raleigh, NC

Scheikl M (2002) Properties of the Glue Line-Microstructure of the Glue Line. In COST Action E13: Wood Adhesion and Glued Products, Working Group 1: Wood Adhesives, State of the Art Report, pp 111-119

Scholz G, Krause A, Militz H (2009) Capillary water uptake and mechanical properties of wax soaked Scots pine. In Proceedings of the Fourth European Conference on Wood Modification, Stockholm, pp 209-212

Sernek M, Resnik J, Kamke FA (1999) Penetration of liquid ureaformaldehyde adhesive into beech wood. Wood Fiber Sci 31(1):41–48

Sernek M (2002) Comparative analysis of inactivated wood surfaces. PhD dissertation, Virginia Polytechnic Institute and State University, Blacksburg, VA, 179 pp

Sernek M, Boonstra M, Pizzi A, Despres A, Gérardin P (2008) Bonding performance of heat treated wood with structural adhesives. Holz Roh Werkst 66:173-180

Shaler SM, Keane DT, Wang H, Mott L, Landis E, Holtxman L (1998) Microtomography of cellulosic structures. In: TAPPI Procceedings, Process and Product, date, city, pp 89-46

Sint KM, Adamopoulos S, Koch G, Hapla F, Militz H (2013) Impregnation of Bombax ceiba and Bombax insigne wood with a N-methylol melamine compound. Wood Sci Technol 47:43–58

Smith LA, Côté WA (1971) Studies of penetration of phenol-formaldehyde resin into wood cell walls with the SEM and energy-dispersive X-ray analyser. Wood Fiber 3:56-58

Stamm AJ, Seborg RM (1936) Resin treated plywood. Ind Eng Chem 31:897–902

Stamm AJ (1959) The dimensional stability of wood. For Prod J 9(10):375–381

Tjeerdsma BF, Stevens M, Militz H, Van Acker J (2002) Effect of process conditions on moisture content and decay resistance on hydro-thermally treated wood. Holzforsch Holzverw 54:94-99

Treu A, Pilgård, Puttmann S, Krause A, Westin M (2009) Material properties of furfurylated wood for window production. International Research Group on Wood Preservation, IRG/WP 09-40480

Trinh HM (2009) Veneer modification for the production of exterior plywood. Sierke Verlag, Thesis Univ. Göttingen

Troughton G.E Chow S.Z (1968) Evidence for covalent bonding between melamine formaldehyde glue and wood. Part I-Bond degradation. Journal of the Institute of Wood Science. 21: 29-33

van Acker J and Hill C (2003) Proceedings of the first European Conference on wood Modification, Ghent, Belgium, pp 1-414

van der Zee ME, Schipholt NL, Tjeerdsma BF, Brynildesn P, Mohoric I (2007) Glueability and paintability of furfurylated wood (Kebony). In: Hill CAS (ed) Proceedings third European conference on wood modification. Cardiff, pp 231–234

van Oss CJ, Giese Jr RF, Good RJ (1990) Reevaluation of the surface tension components and parameters of polyacetylene from contact angles of liquids. Langmuir 6(11):1711-1713

Vick CB, Rowell RM (1990) Adhesive bonding of acetylated wood. Int J Adhes Adhes 10(4):263–272

Vick CB, Larsson PC, Mahlberg RL, Simonson R, Rowell RM (1993) Structural bonding of acetylated Scandinavian softwoods for exterior lumber laminates. Int J Adhes Adhes 13(3):139–149

Vick CB (1999) Adhesive bonding of wood materials. In: Wood Handbook: Wood as an Engineering Material. U.S. Department of Agriculture, Forest Service, Forest Products Laboratory, Madison, WI, chap. 9

Videlov CL (1989) Biological degradation resistance of pine wood treated with dimethylol compounds. International Research Group on Wood Protection, IRG/WP/3528

Wålinder MEP, Johansson I (2001) Measurement of wood wettability by the Willhelmy method. Part 1. Contamination of probe liquids by extractives. Holzforschung 55:21-32

Wålinder MEP, Bryne LE (2006) Wood adhesion mechanisms: prediction of wood-thermoplastic-water interactions". In: Wood Adhesives 2005. CR Frihart (Ed). Forest Products Society. Proceedings No 7230, Madison WI, US, pp 385-392

Weigenand O, Militz H, Tingaut P, Sèbe G, De Jeso B, Mai C (2007) Penetration of amino-silicone micro- and macro-emulsions into Scots pine sapwood and the effect on water-related properties. Holzforschung 61(1):51-59

Westin M, Lande S, Schneider M (2003) Furfurylation of wood—process, properties and commercial production. In: Proceedings of the First European Conference on Wood Modification, Ghent, Belgium, Van Acker, J. and Hill, C.A.S. (Eds.), pp. 289–306

Westin M, Lande S, Schneider M (2004) Wood furfurylation process and properties of furfurylated wood. International Research Group on Wood Preservation, Ljubljana, Slovenia,IRG Secretary Stockholm, Sweden. 1-13

Woods G (1990) The ICI Polyurethanes Book. New York: John Wiley & Sons, Inc. ISBN 0-471-92658- 2

Xiao Z, Xie Y, Militz H, Mai C (2010) Effect of glutaraldehyde on water related properties of solid wood. Holzforschung 64(4):483–488

Xie Y, Krause A, Mai C, Militz H, Richter K, Urban K, Evans PD (2005) Weathering of wood modified with the N-methylol compound 1,3-dimethylol-4,5-dihydroxyethyleneurea. POLYM DEGRAD STABIL, 89: 189-199

Xie Y, Hill CAS, Xiao Z, Militz H, Mai C (2011) Dynamic water vapour sorption properties of wood treated with glutaraldehyde. Wood Sci Technol 45(1):49–61

Xing C, Riedl B, Cloutier A, Shaler SM (2005) Characterisation of urea-formaldehyde resin penetration into medium density fibreboard fibers. Wood Sci Technol 39:374-384

Zee ME, Beckers EPJ, Militz H (1998) Influence of concentration, catalyst, and temperature on dimensional stability of DMDHEU modified Scots pine. International Research Group on Wood Protection, IRG/WP/98-40119

Żenkiewicz M (2007) Methods for the calculation of surface free energy of solids. J Achiev Mater Manuf Eng 24:137-145

Zhang Y, Jin J, Wang S (2007) Effects of resin and wax on the water uptake behavior of wood strands. Wood Fiber Sci 39(2):271–278

Zheng J, Fox SC, Frazier CE (2004) Rheological wood penetration, and fracture performance studies of PF/pMDI hybrid resins. For Prod J 54(10):74-81

6. Publications

With kind permission of: European Journal of Wood and Wood Products

International Journal of Adhesion and Adhesives

Wood Material Science and Engineering

Journal of Adhesion

Wood Research Journal

Paper 1

Eur. J. Wood Prod.
DOI 10.1007/s00107-012-0620-0

Adhesive bonding of beech wood modified with a phenol formaldehyde compound

Stergios Adamopoulos · Alireza Bastani ·
Patricia Gascón-Garrido · Holger Militz ·
Carsten Mai

Received: 7 December 2011
© The Author(s) 2012. This article is published with open access at Springerlink.com

Abstract Untreated (controls) and phenol–formaldehyde (PF)-modified beech wood (10 and 25 % solid content) were glued with phenol resorcinol formaldehyde (PRF) and polyvinyl acetate (PVAc). Shear strength of PRF-bonded specimens was higher than that of PVAc-bonded ones under dry and wet conditions irrespective of the pre-treatment. Under dry conditions, only PVAc-bonded specimens exhibited reduction in shear strength due to PF-modification with 25 % PF concentration as compared to the controls. PF treated wood provided inferior bonding under wet conditions with the exception of 25 % PF concentration specimens glued with PRF adhesive. Modification with PF resulted in a decrease of adhesive penetration into the porous network of interconnected cells, especially at 25 % PF concentration.

Verklebung von mit einer Phenol-Formaldehyd-Verbindung modifiziertem Buchenholz

1 Introduction

Wood modification is a well-established technology to improve biological durability, dimensional stability,

S. Adamopoulos · A. Bastani · P. Gascón-Garrido · H. Militz ·
C. Mai (✉)
Wood Biology and Wood Products, Burckhardt Institute,
Georg-August-University Göttingen, Büsgenweg 4,
37077 Göttingen, Germany
e-mail: cmai@gwdg.de

S. Adamopoulos
Department of Forestry and Management of Natural
Environment, Technological Educational Institute
of Larissa, 431 00 Karditsa, Greece

hardness and weathering resistance of wood (Hill 2006). Besides these advantages, modification can alter the adhesive strength due to changes in chemical, physical and structural properties of wood. The less polar and less porous modified wood surfaces may lead to reduced adhesion due to poorer adhesive wetting of the wood and fewer chemical bonds between the two surfaces (Hunt et al. 2007). On the other hand, improved bonding performance can be achieved as the improved dimensional stability of modified wood results in less shrinking and swelling stresses on the cured adhesive bond (Sernek et al. 2008).

The great variety of wood adhesives, species, and modification methods (chemical, thermal) makes the bonding behaviour of modified wood a complex subject. Acetylation of wood has been shown to affect bonding strength depending on the type of adhesive (Vick and Rowell 1990; Vick et al. 1993; Frihart et al. 2004). The additional hydroxyl groups being available for hydrogen-bonding with the adhesive provided by oxide modifications with butylene and propylene oxide did not give superior adhesion over wood that was acetylated (Brandon et al. 2005). Furfurylated wood could be glued satisfactorily with two different gluing systems even though the percentage glue line failure increased under wet conditions (van der Zee et al. 2007). The poor bonding properties of silicone modified wood were improved by using hydroxymethylated resorcinol as a coupling agent (Kurt et al. 2008). Heat treatment of wood was also reported to affect the glueability of wood in several ways depending on the adhesive type used (Boonstra et al. 1998; Sernek et al. 2008; Sahin Kol et al. 2009).

Impregnation of wood with water soluble, low molecular weight phenol–formaldehyde (PF) resin systems has been researched since the 1930s with main purpose to increase the dimensional stability of solid wood as well as

of the commercial plywood composites Impreg and Compreg (Stamm 1959; Hill 2006; Gabrielli and Kamke 2010). Phenol–formaldehyde resins were shown to penetrate and bulk the cell wall (Stamm and Seborg 1936; Rowell and Banks 1985). Ohmae et al. (2002) obtained anti-shrink efficiencies (ASE) up to 74 at 30 % weight percent gain (WPG) using a low molecular weight PF due to both bulking and cross-linking of the cell wall. PF modification has also been shown to enhance the resistance against white rot and brown rot fungi (Ryu et al. 1991). The effectiveness of modification was shown to depend on the penetration of phenol formaldehyde resins into wood cell walls (Ryu et al. 1993; Furuno et al. 2004).

With new modification processes becoming commercially available and increasing use of modified wood for exterior and interior applications, it is important to continuously update our knowledge on adhesive bonding of modified wood. PF modification has a good chance of commercialisation due to the enhanced wood properties, the relatively low price of PF resin and the relatively easy feasibility of the process. The aim of this study was to evaluate the bonding performance of phenol formaldehyde modified wood glued with two adhesive systems, phenol resorcinol formaldehyde and polyvinyl acetate. The tensile shear bond strength and penetration of adhesives into the porous wood structure were examined.

2 Materials and methods

2.1 Treatment

Defect-free assembly wood blocks of beech (*Fagus sylvatica* L.) with dimensions 600 (longitudinal) × 130 × 6 mm^3 and average density of 680 kg m^{-3} at 12 % moisture content were used for the study. The angle between the growth rings and the surface to be bonded was kept between 30° and 90°. The wood blocks were treated with phenol formaldehyde (PF) with sodium hydroxide added in an aqueous solution with two different concentrations: 10 and 25 % based on the solid content of PF. PF was provided by Cytec Surface Specialties GmbH & Co. KG (Wiesbaden, Germany) with average molecular weight 408 g mol^{-1} (determined by gel permeation chromatography using polystyrene standards), specific gravity 1.15 g cm^{-3}, viscosity 2.760 mPa s at 23 °C, and 78 % solid content. The treatment was conducted in a stainless steel vessel involving a vacuum (100 mbar for 1 h) and a pressure phase (12 bar for 2 h). The wood blocks were then cured gradually from 40 to 120 °C in an oven for a total duration of 11 days. After curing, the blocks were conditioned at 20 °C and 65 % RH in a climate chamber to reach equilibrium moisture content.

2.2 Gluing

Phenol resorcinol formaldehyde PRF (Prefere 4040 with 20 % hardener; Dynea GmbH, Erkner, Germany) and polyvinyl acetate PVAc (Ponal Super 3 with 15 % D4 hardener; Henkel AG & Co., Düsseldorf, Germany) were spread uniformly on the wood blocks by hand brushing. Pressing of the assemblies was carried out for 120 min in a hydraulic press at room temperature and at a pressure of 1.5 N mm^{-2}. Block shear specimens were cut from the assemblies and randomly assigned to either the dry or wet shear tests. Before testing all specimens were conditioned at 20 °C and 65 % RH in a climate chamber for 7 days. Specimens for the wet shear test were further soaked in water at 15 ± 5 °C for 4 days and tested in the wet state. The measurement of longitudinal shear strength was carried out in a Zwick/Roell Z010 universal testing machine according to the European Standard (1992) EN 302-1. Ten replicates were used for each adhesive and treatment.

2.3 Adhesive penetration

To observe the adhesive penetration into the porous wood structure, 20–30 μm thick sections exposing a bondline with a cross-sectional surface at various longitudinal positions of the specimens were cut on a Reichert-Jung sliding microtome. The sections were placed on glass slides, unstained for PRF bondlines and after staining with a pipette drop of 0.5 % safranin O solution for PVAc bondlines. PVAc sections were examined under an Eclipse 50i fluorescence microscope with appropriate filter sets, equipped with a Sight DS-5M-L1 digital camera (both Nikon, Düsseldorf, Germany), while conventional light microscopy was applied for PRF sections by using the same microscope. Penetration depth of the adhesive was measured on 25 positions along 130 mm of bondline for each section.

3 Results

Table 1 presents tensile shear strength and wood failure data results obtained for each adhesive. Under both dry and wet condition, PRF adhesive produced higher shear strengths and wood failure percentages than PVAc irrespective of the pre-treatment. In the dry test, bonding strength remained unchanged after modification with PF except for the specimens modified with 25 % concentration when glued with PVAc; these showed a significantly lower strength of 6.13 N mm^{-2} (ANOVA and Tukey HSD test, $P = 5$ %). This combination also gave the lowest wood failure percentage (40 %), while wood failure percentages of 80–90 % for all other cases verify the

 Springer

acceptable bonds produced by the PF-treated wood under dry conditions. Under wet conditions, PVAc bonds weakened considerably giving almost zero shear strength values $(0.15–0.60 \text{ N mm}^{-2})$ for PF-modified wood and correspondingly very little wood failure (10 %). Exposure to water-soaking almost halved the bonding strength of

Table 1 Bonding strength of beech wood modified with phenol formaldehyde (PF)
Tab. 1 Klebefestigkeit von mit Phenol-Formaldehyd (PF) behandeltem Buchenholz

Adhesive type/PF solid content	Tensile shear strength (N/mm²)				t	Mean wood failure (%)	
	Dry condition		Wet condition			Dry condition	Wet condition
	Mean	±SD	Mean	±SD			
PVAc/Control	8.79^a	3.44	3.73^a	1.53	3.880*	90	20
PVAc/10 %	8.37^{ab}	1.98	0.60^b	0.27	12.267*	80	10
PVAc/25 %	6.13^b	1.31	0.15^b	0.08	10.980*	40	10
F	4.273*		36.214*				
PRF/Control	11.17	2.01	7.51^{ab}	1.25	4.822*	90	60
PRF/10 %	10.23	2.12	5.97^b	1.58	5.040*	80	40
RPF/25 %	9.92	3.01	8.58^a	2.02	1.170^{ns}	80	70
F	0.686^{ns}		5.933*				

Values followed by a different letter within a column are statistically different at $P = 5$ % (ANOVA and Tukey HSD test)

* Differences statistically significant at $P = 5$ %

^{ns} Differences not statistically significant

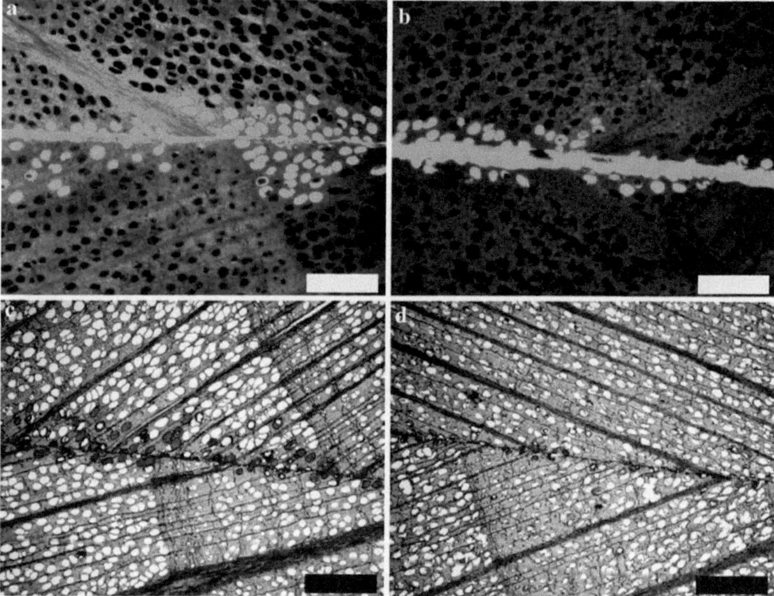

Fig. 1 Transverse view of horizontal bondlines in beech wood modified with phenol formaldehyde in 10 % (**a, c**) and 25 % concentration (**b, d**) using fluorescence light for PVAc (**a, b**) and visible light for PRF adhesive (**c, d**). *Scale bars*: 500 μm
Abb. 1 Querschnitt der horizontalen Klebstoffschicht von Buchenholz, das mit 10%iger (**a, c**) und 25%iger Phenol-Formaldehyd-Lösung modifiziert wurde, wobei zur Detektion von PVAc Fluoreszenzlicht (**a, b**) und von PRF-Klebstoff sichtbares Licht (**c, d**) verwendet wurde. Maßstabsskala: 500 μm

10 % PF-modified and PRF-bonded samples but had no statistically significant effect for 25 % PF pre-treatment (t test, $P = 5$ %). Differences in wood failure percentages between the dry and wet condition were in agreement with the shear strength data for the PRF adhesive. Improved adhesion with PRF adhesive in wet conditions could be obtained with increasing PF concentration. The 25 % PF treated samples showed higher wood failure (70 %) and strength (8.58 N mm^{-2}) than the 10 % PF-treated ones (respective values were 40 % and 5.97 N mm^{-2}; for shear strength, differences were statistically significant using Tukey's analysis at $P = 5$ %). However, shear strength of modified wood in each concentration did not differ significantly when compared to unmodified wood (ANOVA and Tukey HSD test, $P = 5$ %).

For all treatments and adhesive types, a discontinuous interphase (e.g. volume in which both wood cells and adhesive are present) was seen further away from the bondline (Fig. 1). Penetration was dominated by flow through the vessels and appeared either as total or partial filling of the lumens. The observations of this study did not reveal any evidence of adhesive flow in the rays while penetration of adhesive into the fibres and axial parenchyma was limited to only very few cells (1–3) near the bondline. PVAc penetration in the vessels could be observed up to a distance of approximately 200 μm from the contact area between the wood surface and the adhesive for the control and the 10 % PF-modified specimen, while in the case of pre-treatment with 25 % PF penetration depth was limited up to approximately 130 μm. Similar observations were made for the PRF adhesive in 25 % PF-modified specimens presenting the lower penetration depth (approx. 160 μm). Control and 10 % PF-modified samples showed comparable average penetration values of approximately 240 and 220 μm, respectively. Deposition of the PF resin into the cell wall and lumens as well as its hydrophobation effect (Furuno et al. 2004) affected adhesive penetration at 25 % PF concentration but resulted in significant lower adhesion only for PVAc.

4 Conclusion

The bonding performance of PF modified beech wood using PVAc and PRF adhesives was examined in this study. PRF adhesive causes higher shear strength than PVAc under both dry and wet condition. PF modified wood can be bonded satisfactorily with the two adhesive systems under dry conditions with the exception of wood modified with 25 % PF solution and PVAc adhesive. PVAc adhesive should be avoided for gluing of PF modified wood to be used for exterior applications as water-soaking results in almost zero shear bond strength and very little wood failure. In exterior applications, the structural adhesive PRF and wood modified with high PF concentration was the best combination to achieve optimal bonding performance. For both adhesives, the penetration into the porous wood structure is the lowest at highest degree of PF modification.

References

Boonstra MJ, Tjeerdsma BF, Groeneveld HAC (1998) Thermal modification of non-durable wood species. Part 1, The Plato technology: thermal modification of wood. International Research Group on Wood Preservation, Document no. IRG/WP 98-40123

Brandon R, Ibach RE, Frihart CR (2005) Effects of chemically modified wood on bond durability. In: Frihart CR (ed) Wood Adhesives 2005. Forest Products Research Society, San Diego, pp 111–114

European Standard (1992) EN 302-1: adhesives for load bearing timber structures—test methods—Part 1: determination of bond strength in longitudinal tensile shear

Frihart CR, Brandon R, Ibach RE (2004) Selectivity of bonding for modified wood. P Adhes Soc 27(1):329–331

Furuno T, Imamura Y, Kajita H (2004) The modification of wood by treatment with low molecular weight phenol-formaldehyde resin: a properties enhancement with neutralized phenolic-resin and resin penetration into wood cell walls. Wood Sci Technol 37:349–361

Gabrielli CP, Kamke FA (2010) Phenol-formaldehyde impregnation of densified wood for improved dimensional stability. Wood Sci Technol 44:95–104

Hill CAS (2006) Wood modification: chemical, thermal and other processes. Wiley, Chichester

Hunt CG, Brandon R, Ibach RE, Frihart CR (2007) What does bonding to modified wood tell us about adhesion? In: Proceedings 5th COST E34 international workshop: bonding of modified wood. Bled, Slovenia, pp 47–56

Kurt R, Mai C, Krause A, Militz H (2008) Hydroxymethylated resorcinol (HMR) priming agent for improved bondability of silicone modified wood glued with a polyvinyl acetate adhesive. Holz Roh Werkst 66:305–307

Ohmae K, Minato K, Nonmoto M (2002) The analysis of dimensional changes due to chemical treatments and water soaking for hinoki (*Chamaecyparis obtusa*) wood. Holzforschung 56:98–102

Rowell RM, Banks WB (1985) Water repellency and dimensional stability of wood. USDA General Technical Report FPL-50:1–24

Ryu JY, Takahashi M, Imamura Y, Sato T (1991) Biological resistance of phenol-resin treated wood. Mokuzai Gakkaishi 37(9):852–858

Ryu JY, Imamura Y, Takahashi M, Kajita H (1993) Effects of molecular weight and some other properties of resins on the biological resistance of phenolic resin treated wood. Mokuzai Gakkaishi 39(4):486–492

Sahin Kol H, Özbay G, Altun S (2009) Shear strength of heat-treated tali (*Erythrophleum ivorense*) and iroko (*Chlorophora excelsa*) woods, bonded with various adhesives. Biores 4(4):1545–1554

Sernek M, Boonstra M, Pizzi A, Despres A, Gérardin P (2008) Bonding performance of heat treated wood with structural adhesives. Holz Roh Werkst 66:173–180

Stamm AJ (1959) The dimensional stability of wood. For Prod J 9(10):375–381

Stamm AJ, Seborg RM (1936) Resin treated plywood. Ind Eng Chem 31:897–902

van der Zee ME, Schipholt NL, Tjeerdsma BF, Brynildesn P, Mohoric I (2007) Glueability and paintability of furfurylated wood (Kebony). In: Hill CAS (ed) Proceedings Third European Conference on Wood Modification. Cardiff, UK, pp 231–234

Vick CB, Rowell RM (1990) Adhesive bonding of acetylated wood. Int J Adhes Adhes 10(4):263–272

Vick CB, Larsson PC, Mahlberg RL, Simonson R, Rowell RM (1993) Structural bonding of acetylated Scandinavian softwoods for exterior lumber laminates. Int J Adhes Adhes 13(3):139–149

Paper 2

Eur. J. Wood Prod. (2015) 73:627–634
DOI 10.1007/s00107-015-0919-8

ORIGINAL

Water uptake and wetting behaviour of furfurylated, *N*-methylol melamine modified and heat-treated wood

Alireza Bastani[1] · Stergios Adamopoulos[2] · Holger Militz[1]

Received: 26 April 2014 / Published online: 6 May 2015
© Springer-Verlag Berlin Heidelberg 2015

Abstract This study reports on the water uptake (WU) and wetting properties of different modified wood materials; furfurylated and *N*-methylol melamine (NMM) modified Scots pine, and heat-treated (Vacu^3 method) Scots pine and beech. All modifications caused a substantial reduction in WU in the longitudinal, tangential and radial directions both after short (24 h) and long contact times (168, 336 h) with a saturated sponge. The water uptake coefficient (w_t) was reduced by approximately 71–89 % in furfurylated wood, with the higher weight percent gain (WPG) providing a slightly greater reduction. The reduction in WU was not found to depend on the NMM solid content. The NMM treatment had the maximum effect on the reduction of tangential w_t by 80–84 % and was much smaller in the longitudinal direction (31–68 %). The treatment temperature of 195 °C gave lower WU values than treatment at 210 °C, and the only exception was the radial direction of Scots pine. The longitudinal w_t of heat-treated beech represented the highest reduction by 81–89 %, while radial w_t was less affected in both species. Sessile drop apparent contact angles for water and diidomethane and corresponding surface energies on planed tangential and radial wood surfaces revealed an increased hydrophobicity and reduced polarity of modified wood. Furfurylated and NMM modified tangential surfaces had a higher increase of apparent contact angles than the radial surfaces but this was not observed in the case of heat treatment. Heat-treated wood showed reduced wetting of surfaces only with water. Apparent contact angles did neither differ with treatment temperature nor with the NMM resin load. The disperse component of surface energy was slightly increased by 20 % maximum in modified wood, while the polar components showed a dramatic decrease by −30 to −90 % with no major differences among treatments and intensities, and between surfaces. The results provide a better understanding of the hygroscopic behaviour of modified wood, which might be useful to predict its adhesion with various polymers such as glues, coatings and paints.

✉ Alireza Bastani
abastan@gwdg.de

[1] Wood Biology and Wood Products, Burckhardt Institute,
Georg-August-University Göttingen, Büsgenweg 4,
37077 Göttingen, Germany

[2] Department of Forestry and Wood Technology,
Linnaeus University, 351 95 Växjo, Sweden

1 Introduction

Modification treatments have been studied and applied widely with the aim to improve various technological properties of wood and wood-based products such as dimensional stability, weathering, and durability (Hill 2006). Bulk treatments (e.g. with resins, waxes) by blocking pathways for the water flow and due to the hydrophobic character of the chemicals as well as chemical (e.g. acetylation, furfurylation) or thermal modification by changing the chemical character of wood cell walls due to decreasing of e.g. hydroxyl groups result in reduced water uptake and equilibrium moisture content of wood (Militz 2002; Gsöls et al. 2003; Pétrissans et al. 2003, Mai and Militz 2004; Weigenand et al. 2007; Zhang et al. 2007; Xiao et al. 2010). Although the purpose of the treatments has been achieved, leading thereafter to improved dimensional stability and durability of wood, issues with regard to processing of modified wood such as coating, painting,

and gluing might arise due to the more hydrophobic wood surfaces.

The study of wood surface characteristics is of vital interest to understand the effect of various modifications on the adhesion between wood and various polymers such as coatings and glues. According to the wetting or adsorption theory, for example, the less polar and in some cases less porous modified wood surfaces may lead to reduced adhesion due to poorer adhesive wetting of the wood (Hunt et al. 2007).

The wettability of wood can be defined by parameters which describe the molecular polar or non-polar interactions between liquids and solids such as contact angles, surface free energy and work of adhesion (Mantanis and Young 1997; de Meijer et al. 2000; Wålinder and Bryne 2006). Contact angle analysis is a useful tool for estimations of solid–liquid interfacial forces and is performed by using the Wilhelmy, the rising height, and the sessile drop methods (Pétrissans et al. 2003; Bryne and Wålinder 2010). A commonly used technique for investigations on wood is the sessile drop method, which is based on the observation of the profile of a drop deposited on a wood surface (Neumann and Spelt 1996). Limitations associated with the technique include bulk sorption, roughness and porosity of wood, and probe liquid contamination as a result of the wood extractives. The term "apparent contact angle" is preferred because it is a more accurate expression for the measurement of contact angle due to the mentioned finiteness (Bryne and Wålinder 2010).

It is also well-known that swelling and shrinkage of wood has a major effect on the performance of synthetic polymers applied to its surface. Modification techniques reduce the hydrosensitivity of wood substrate and considerably improve its dimensional stability, thus resulting in less wood-polymer debonding (Stamm 1964; Hill 2006). Besides the standard methods used for determining water uptake and water vapour sorption, e.g. submersion in water or moisture sorption tests (Donath et al. 2006; Xie et al. 2011; Ghosh et al. 2013; Pries et al. 2013), the hygroscopic behaviour of modified wood is also evaluated on the basis of water uptake of samples in contact with a saturated sponge (Ghosh et al. 2009; Scholz et al. 2009; Johansson and Kifetew 2010; Xiao et al. 2010). Due to the porous character of wood, the water flows through capillary force in the wood, and also diffuses into the cell walls, and consequently the method used provides estimations of its water uptake behaviour in exterior use as well as of the localisation of chemicals in different morphological regions of modified woody tissues, which may block the fluid flow path (Rowell and Banks 1985).

The objective of this work was to study the wetting phenomena and water uptake of different modified wood materials. As several wood modification technologies are

being commercially exploited today and modified wood is increasingly used in combination with other materials such as adhesives, varnishes, paints and coatings, such information could be used to predict the performance of modified wood-polymer systems.

2 Materials and methods

2.1 Wood material and modification methods

The modified wood materials studied were furfurylated Scots pine (*Pinus sylvestris* L.), melamine treated Scots pine, and heat treated Scots pine and beech (*Fagus sylvatica* L.). For the modifications, Scots pine (sapwood) and beech boards were used with dimensions $1400 \times 100 \times 30$ mm^3 (L $\times$ W $\times$ T).

Furfurylation was done by Kebony ASA (Norway) within an industrial process with two types of furfuryl alcohol, which are industrially known as Kebony FA 40 and FA 70, with a 65 and 75 % WPG, respectively (Mai 2010).

For the melamine treatment, the *N*-methylol melamine (NMM) resin Madurit MW840/75WA (Ineos Melamines GmbH, Frankfurt, Germany) was supplied as an aqueous stock solution with a solid content of approx. 75 %. Solutions of 10, 20 and 30 % NMM solid content were prepared and impregnation of wood boards was done in a stainless steel vessel using a full cell process, which included an initial vacuum phase of 100 mbar for 1 h and a pressure phase of 12 bar for 2 h. Curing of the NMM resin took place at 120 °C for 1 week in a drying chamber. Heat-treatment was carried with the industrial scale vacuum-press dewatering method (Vacu^3) at 195 and 210 °C by Timura Holzmanufaktur GmbH (Germany). In this method, treatment temperature devolves the wood using heating plates enabling a very efficient heat transfer to the lumber, while due to the high vacuum used by-products are condensed at the dryer's wall and taken out of the system (Niemz 2007). Prior to the testing, all modified wood materials were conditioned at 20 °C and 65 % RH in a climate chamber.

2.2 Water uptake

Water uptake (WU) was measured along the longitudinal direction on samples with dimensions of $20 \times 20 \times 200$ mm^3 (R $\times$ T $\times$ L) and along the tangential and radial directions on samples measuring $40 \times 40 \times 40$ mm^3 (R $\times$ T $\times$ L) according to DIN 52617 (1987). Ten replicates per treatment and direction were used. The samples were sealed with a water resistant coating along all faces except for two (transverse, radial or tangential) to test each time the longitudinal, tangential and radial water uptake.

 Springer

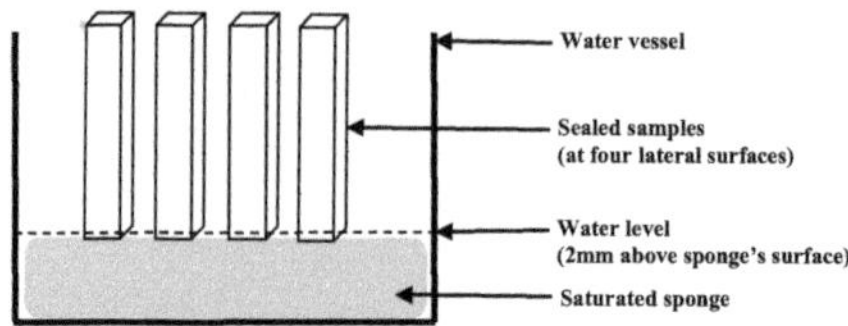

Fig. 1 Experimental set-up for the capillary water uptake test showing a water vessel with wood samples in contact with a saturated sponge. Note that the water level is kept 2 mm above the sponge surface and the samples are sealed with a water resistant coating along all faces except two

The samples were then placed in a water vessel in contact with a saturated sponge to absorb water through the open faces (Fig. 1). The water level was kept 2 mm above the sponge surface, such that a non-sealed surface of the samples was only partially immersed in water to assess the intensity of water absorption. WU was measured after specific time intervals of 1, 2, 3, 4, 8, 12, 24, 48, 72, 96, 120, 168 and 336 h. In accordance with DIN 52617 (1987), the water uptake coefficient was calculated for every direction by using the equation, $w_t = \Delta W_t t^{-0.5}$, where w_t is the water uptake coefficient (kg m^{-2} h$^{-0.5}$), ΔW_t is the difference in mass of water uptake per unit area of sample surface (kg m^{-2}) between start and time t; and t is the measuring time (h).

2.3 Contact angle and surface energy

The apparent sessile drop contact angles were measured with the G 10 device (Krüss GmbH, Hamburg, Germany) and DSA 1 software on planed tangential and radial surfaces of wood samples. The probe liquids included water and diiodomethane (dosing volume of 10 µl), and the apparent contact angle value was taken 5 s after droplet deposition. Preliminary tests showed that this period of time was necessary in order to obtain droplet stabilization in both unmodified and modified wood. Furthermore, it has been reported that contact angles measured about 5 s after depositing the water drop on the wood surface minimise the risk of solvent contamination with wood extractives (Wålinder and Johansson 2001). For each sample, 20 representative measurements were performed at different spots to minimise the effects of structural and chemical variations of the wood samples.

There are various methods described in the literature for the calculation of surface energy of solids from contact angle data, mainly using the Young equation as a basis (Żenkiewicz 2007). In the individual methods, various assumptions have been made in finding mathematical relationships to describe quantitatively the phenomena of interfacial interactions in the systems of various liquids and polymeric materials. There appears to be no consensus on the use of the different approaches for surface characterization (Gindl et al. 2001), and thus the surface energy values depend on the method of measurement and analytical computation. In this work, the geometric mean approach developed by Owens–Wendt (1969) was used, which is a commonly used method for calculating the surface energy of wood. The mean values of contact angle measurements were used to calculate the corresponding surface energy by following the Owens–Wendt approach on the basis of Young's equation, $\gamma_S = \gamma_{SL} + \gamma_L \cos \theta$, where γ_S is the surface energy of the solid in mN m^{-1}, γ_{SL} is the interfacial energy between solid and liquid, γ_L is the surface tension of the liquid, and θ is the apparent contact angle. In this approach, the total surface energy γ^T is considered as a sum of a disperse or non-polar component γ^D and a polar component γ^P as: $\gamma^T = \gamma^D + \gamma^P$. A minimum of two measuring liquids is required to determine the dispersed and polar components of surface tension, usually water and diiodomethane. However, the more liquids that are used the more reliable the results become. The surface tension components γ^D and γ^P for the probe liquids are respectively, 21.8 and 51 mN m^{-1} for water, and 50.8 and 0 mN m^{-1} for diiodomethane (van Oss et al. 1990).

3 Results and discussion

3.1 Water uptake

The faster water transport through the axially oriented tracheids of Scots pine and vessel elements of beech caused the highest water uptake (WU) in the longitudinal direction compared to the lateral ones for all modified and unmodified samples (Fig. 2; Table 1). The slightly higher radial WU as compared to the tangential one can be attributed to the presence of rays with large lumens and thin cell walls and open pit structures running along the radial direction, which facilitate the flow of water inside the wood structure (Weigenand et al. 2007; Xiao et al. 2010).

WU was reduced substantially in all directions both after short (24 h) and long submersion times (168, 336 h) by the modifications (Fig. 2; Table 1). For furfurylated wood, the treatment with higher uptake of furfuryl alcohol (FA 70) caused a slightly greater reduction in water uptake in all directions. This reduction effect for higher levels of furfurylation was also shown by Treu et al. (2009) for beech wood. The water uptake coefficient (w_t) was reduced by approximately 84–89 % after 24 h; the values were even high (71–82 %) after 168 and 336 h (Table 1).

As previously reported (Sint 2010), the reduction in water uptake did not follow a certain pattern based on the concentration of the NMM treatment solution. While no

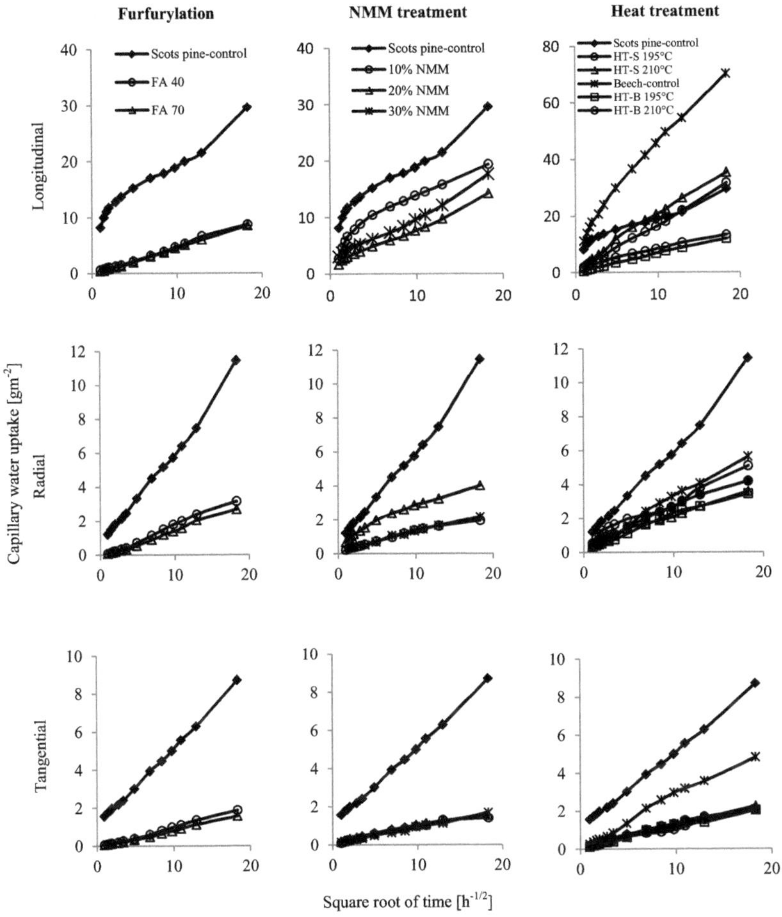

Fig. 2 Longitudinal, radial and tangential capillary water uptake of furfurylated Scots pine (FA 40 and 70), NMM modified Scots pine (10, 20 and 30 % NMM), and heat-treated Scots pine (HT-S) and beech (HT-B)

differences were noted in the tangential direction among the NMM concentrations, 20 % NMM performed better than the others in the longitudinal direction but it was inferior to 10 and 30 % NMM in the radial direction (Fig. 2). The NMM treatment had the maximum effect on the reduction of tangential w_t in all concentrations (80–84 %) already after 24 h (Table 1). These values were comparable to those of radial w_t with the exception of 20 % NMM (41–65 %). In contrast, NMM treatment reduced the longitudinal w_t after 24 h by only 31–68 %. In the case of 20

and 30 % NMM, the values even declined from 59–68 % to 40–52 % after 336 h. The lower water uptake of NMM treated wood can be explained by the penetration of NMM resin in different morphological regions of wood tissues and cell walls, and by resulting cell wall bulking (Mahnert et al. 2013; Sint et al. 2013; Kielmann et al. 2014). Thus, the available sites in the cell wall to absorb water molecules are reduced. However, the reduction in the velocity of liquid water uptake in the melamine treated wood (Fig. 2) should be attributed mainly to the occlusion of the major

Table 1 Water uptake coefficient (w_t) of furfurylated Scots pine (FA 40 and 70), NMM modified Scots pine (10, 20 and 30 % NMM), and heat-treated Scots pine (HT-S) and beech (HT-B) in longitudinal, radial and tangential directions after 24, 168 and 336 h

Treatment	w_t (kg m^{-2} h$^{-0.5}$)								
	Longitudinal			Radial			Tangential		
	24 h	168 h	336 h	24 h	168 h	336 h	24 h	168 h	336 h
Scots pine-control	74.57	278.64	542.75	16.28	96.68	209.87	14.76	81.51	159.47
Beech-control	146.4	708.26	1,289.14	7.18	52.74	103.38	6.6	46.39	88.53
FA 40	10.36	85.79	159.47	3.42	30.97	57.73	1.85	17.49	34.09
FA 70	9.53	77.63	155.8	2.64	26.3	49.12	1.56	14.51	28.59
10 % NMM	51.14	205.02	355.78	3.52	20.99	35.56	2.68	16.45	25.66
20 % NMM	23.66	126.36	261.75	9.58	41.86	73.5	2.98	16.2	28.22
30 % NMM	30.26	157.85	324.25	3.32	21.38	39.4	2.39	14.64	30.97
HT-S 195 °C	43.47	284.34	581.24	9.63	49.24	93.29	3.32	19.82	38.49
HT-S 210 °C	57.65	342.01	648.88	6.84	34.86	65.62	3.86	20.99	41.05
HT-B 195 °C	16.62	113.78	218.67	5.42	34.99	63.05	3.03	18.27	37.39
HT-B 210 °C	26.06	136.46	242.87	7.23	43.93	76.8	3.61	21.9	37.94

penetration pathways for water, e.g. lumens of axial elements (tracheids, vessel elements) and ray cells, by the resin rather than its incorporation in the cell walls.

Heat treatment substantially reduced the WU of wood (Fig. 2), as a result of thermal degradation of wood polymers, mainly destruction and deacetylation of hemicelluloses beginning below 180 °C, and increase of cellulose crystallinity (Militz 2002). Other phenomena such as structural modifications and changes in the chemical structure of lignin, especially when the treatment temperature is above 200 °C, might also play an important role (Alén et al. 2002; Sivonen et al. 2002). Changes in the lignin fraction are mostly registered as depolymerisation of lignin macromolecules, diminishment in the methoxyl content, cleavage of the polysaccharide lignin complexes, and condensation reactions leading to crosslinks within the lignin and possibly between lignin and other wood components (Hill 2006). Changes in the spatial distribution of hydrophobic extractives within the bulk and on the surface of wood after heat treatment (Hakkou et al. 2005; Nuopponen et al. 2003) should also be mentioned. A deviation from the general pattern of reduced WU referred to the higher longitudinal uptake of treated Scots pine than the untreated one at longer submersion times (Fig. 2; Table 1). After the long submersion time of 336 h the increase of longitudinal w_t amounted to 7–20 % while previously the short submersion period (24 h) caused a decrease by 23–42 % (Table 1). The phenomenon has previously been observed for Scots pine sapwood at temperatures of 170, 190, and 210 °C compared to untreated material. Water absorption did not decrease until the heat treatment temperature was 230 °C (Metsä-Kortelainen et al. 2006). With the exception of radial uptake of Scots pine, in all other cases the lower treatment temperature of 195 °C gave better results than the treatment at 210 °C. It should be noted that wood defects occurring during heat treatment such as cracking and broken cell walls might also have influenced the results. For example, a collapse of vessels (Boonstra et al. 2006) could explain the substantial reduction in the longitudinal uptake of heat-treated beech (Fig. 2; Table 1). As can be seen in Table 1, this represented the highest reduction in w_t (81–89 %) by the heat treatment and was followed by the tangential w_t of Scots pine (74–78 %). Radial w_t was less affected in both species, and also differences between short and long submersion times were generally small (Table 1).

3.2 Contact angle

Furfurylation generally resulted in a decreased level of wetting force, i.e., an increased apparent contact angle for both probe liquids (Table 2). The highest increase in contact angles was for tangential surfaces. An exception was seen for the radial surface of furfurylated F40 wood which had slightly lower contact angles for water than the unmodified surface. This was also reported by Bryne and Wålinder (2010) and was attributed to the presence of hygroscopic buffering salt agents used in the furfurylation process. The presence of small amounts of unreacted furfuryl alcohol and other by-products of the process might also be responsible for the lower contact angle values (Englund et al. 2009). It should also be noted that the process additives might potentially cause probe liquid contamination and therefore might affect the surface energy properties of furfurylated wood.

A significant increase of hydrophobicity was caused by the NMM treatment and it was independent from the NMM resin load (Table 2). As also observed for furfurylation, the NMM modified tangential surfaces had a higher increase of contact angles than the radial surfaces for both probe

Table 2 Mean values of apparent contact angles for water and diiodomethane on radial and tangential surfaces of furfurylated Scots pine (FA 40 and 70), NMM modified Scots pine (10, 20 and 30 % NMM), and heat-treated Scots pine (HT-S) and beech (HT-B)

Treatment	Contact angle (°)			
	Water		Diiodomethane	
	Radial	Tangential	Radial	Tangential
Scots pine-control	69.2 (21.8)	47.0 (21.0)	20.5 (5.8)	14.9 (7.4)
Beech-control	72.1 (9.1)	77.9 (14.6)	28.1 (3.9)	28.4 (5.9)
FA 40	61.3 (22.7)	68.0 (28.3)	24.6 (5.7)	22.0 (6.7)
FA 70	79.7 (11.7)	79.8 (16.4)	24.2 (6.9)	18.7 (6.6)
10 % NMM	83.9 (9.8)	76.2 (9.2)	25.1 (5.7)	21.4 (4.1)
20 % NMM	84.6 (9.4)	74.6 (12.7)	21.6 (3.6)	18.7 (4.2)
30 % NMM	84.2 (13.8)	66.7 (16.6)	23.9 (4.3)	19.1 (4.4)
HT-S 195 °C	86.1 (11.9)	81.7 (15.5)	30.2 (4.0)	14.6 (7.6)
HT-S 210 °C	81.4 (9.7)	79.8 (13.1)	20.3 (6.4)	14.5 (8.9)
HT-B 195 °C	80.5 (9.9)	73.2 (12.6)	28.5 (4.8)	19.4 (3.9)
HT-B 210 °C	83.1 (13.2)	79.5 (14.3)	26.3 (5.9)	20.7 (3.9)

Standard deviations in parenthesis

liquids. However, the radial surfaces had higher contact angles and thus a lower wetting force.

Heat treatment also affected negatively the wetting of wood surfaces but only for water (Table 2). The migration of extractives at the surface after the heat treatment and the risks associated with the contamination of this probe liquid (Bryne and Wålinder 2010) might be the potential reasons for this result. Specifically, the exudation of mobile hydrophobic substances from the interior to the wood surface causes a minimization of system energy by exposing a more non-polar hydrophobic phase towards the air (Englund et al. 2009). For both species, no major differences were found in contact angles with the treatment temperature (195 and 210 °C) and between the radial and tangential surfaces. However, the contact angles on the tangential surfaces of beech did not change much after heat treatment. In contrast, contact angles for diiodomethane mostly remained unchanged or even decreased in heat-treated wood (e.g. tangential surfaces of beech for both treatment temperatures). A positive effect was only noted for the radial surface of Scots pine treated at 195 °C.

Most literature studies showed a significant increase in wood hydrophobicity after heat treatment. After lab scale heat treatment in the temperature range between 160 and 260 °C, the wood of four European species (pine, spruce, beech and poplar) had a pronounced hydrophobic character with contact angle reaching an average value of 90° (Hakkou et al. 2005). The observations confirmed a previous study using wood of these species treated industrially at 240 °C under inert atmosphere (Pétrissans et al. 2003). Heat treatment of North American white ash and soft maple under gas atmosphere without the presence of oxygen up to 215 °C decreased wood wettability with no differences between the contact angles on tangential and radial surfaces (Kocaefe et al. 2008). The lignin rich surfaces of thermally treated spruce wood were not found to be hydrophobic based on their contact angle data (Bryne and Wålinder 2010).

3.3 Surface energy

All the modifications had a clear impact on the surface energy characteristics leading to a non-polar character of radial and tangential surfaces (Table 3). This is due to an increase in hydrophobicity of wood after modification as a result of chemical changes, which reduce its surface energy (Hill 2006). Thus, the total surface energy was mostly determined by the disperse component of the surface energy. The disperse component of surface energy was slightly increased in modified wood. This effect was more evident for the tangential surfaces (increase up to 15–16 % in heat-treated Scots pine) than the radial surfaces (increase up to 6–7 % in 20 % NMM treated wood, and heat-treated Scots pine and beech at 210 °C). The differences in tangential and radial surface energies may be attributed to differences in wood extractive composition and the diversities in anatomical features in these directions (Nussbaum 1999). The polar components, on the other hand, showed a dramatic decrease ranging from −30 to −90 % with no major differences among treatments and intensities, and between surfaces (radial, tangential). An exception to the above mentioned patterns was observed for disperse and polar components of the radial surface of the F40 modified wood and for the polar component of the tangential surface of the heat-treated beech at 195 °C. The very low surface polarity of modified wood is expected to prevent the formation of attractive polar forces, impair wetting, and accordingly affect negatively the adhesion of glues and paints.

 Springer

Table 3 Mean values of disperse (γ^D) and polar (γ^P) components (mN m^{-1}) of the total surface energy (γ^T) on radial and tangential surfaces of furfurylated Scots pine (FA 40 and 70), NMM modified Scots pine (10, 20 and 30 % NMM), and heat-treated Scots pine (HT-S) and beech (HT-B)

Treatment	Radial			Tangential		
	γ^D	γ^P	γ^T	γ^D	γ^P	γ^T
Scots pine-control	43.1 (1.6)	6.8 (1.5)	49.9 (3.1)	40.7 (1.9)	19.3 (2.5)	60.1 (4.4)
Beech-control	40.8 (1.0)	6.1 (0.6)	47.0 (1.6)	41.8 (1.5)	3.7 (0.7)	45.6 (2.3)
FA 40	40.2 (1.5)	11.5 (2.0)	51.7 (3.5)	42.4 (1.8)	7.5 (2.0)	49.9 (3.9)
FA 70	43.9 (1.8)	2.8 (0.5)	46.7 (2.4)	45.9 (1.8)	2.4 (0.7)	48.3 (2.5)
10 % NMM	44.5 (1.5)	1.6 (0.3)	46.1 (1.8)	44.2 (1.0)	3.8 (0.5)	48.1 (1.6)
20 % NMM	45.9 (0.9)	1.3 (0.2)	47.3 (1.2)	44.8 (1.1)	4.3 (0.7)	49.1 (1.8)
30 % NMM	45.0 (1.1)	1.5 (0.4)	46.5 (1.6)	43.0 (1.2)	7.9 (1.2)	51.0 (2.4)
HT-S 195 °C	42.7 (1.0)	1.3 (0.1)	44.1 (1.4)	47.4 (2.0)	1.7 (0.5)	49.2 (2.6)
HT-S 210 °C	45.7 (1.7)	2.0 (0.4)	47.8 (2.1)	47.0 (2.4)	2.2 (0.6)	49.3 (3.0)
HT-B 195 °C	42.3 (1.2)	2.8 (0.4)	45.2 (1.7)	44.3 (1.0)	4.9 (0.7)	49.2 (1.8)
HT-B 210 °C	43.8 (1.5)	1.9 (0.5)	45.7 (2.0)	45.2 (1.1)	2.6 (0.6)	47.8 (1.7)

Standard deviations in parenthesis

4 Conclusion

The changes in the hygroscopic behaviour and wetting properties of wood caused by various modifications have significant consequences, in particular for the interaction and adhesion of polymers (e.g. glues, paints, coatings). The present analysis of water uptake (WU), apparent contact angle data and surface energy characteristics of furfurylated, NMM modified and heat-treated wood lead to the following conclusions:

- All modifications resulted in a considerable reduction of WU in the three main directions of wood even after a short contact time (24 h) with a saturated sponge. For furfurylated wood, the higher WPG of furfuryl alcohol had a slightly better effect. The reduction in WU was not found to depend on the concentration of the NMM treatment solution. The NMM treatment had the maximum effect on the tangential WU and the lowest on the longitudinal one. With the exception of radial uptake of Scots pine, heat treatment at 195 °C gave better results than the treatment at 210 °C. The lowest WU caused by the heat treatment was noted longitudinally for beech.
- Hydrophobicity of wood, as expressed by apparent contact angle data, was increased by the modifications. Some exceptions were observed, mainly for heat-treated wood. The load of furfuryl alcohol and NMM had no effect on contact angles, and so it was for the treating temperatures in the case of heat-treated wood. For both furfurylation and NMM treatment the tangential surfaces showed the highest increase of contact angles, while the temperature had no effect on contact angles between the heat-treated surfaces (radial, tangential).

- Modifications provided radial and tangential surfaces with a non-polar character due to the vast decrease of the polar components of the surface energy and the slight increase of the disperse ones. No major differences in surface polarity were found for different treatments, intensities, and wood surfaces.

References

Alén R, Kotilainen R, Zaman A (2002) Thermochemical behavior of Norway spruce (*Picea abies*) at 180–225 °C. Wood Sci Technol 36:163–171

Boonstra MJ, Rijsdijk JF, Sander C, Kegel E, Tjeerdsma B, Militz H, van Acker J, Stevens M (2006) Microstructural and physical aspects of heat treated wood. Part 2. Hardwoods. Maderas Cienc Tecnol 8:209–217

Bryne LE, Wålinder MEP (2010) Ageing of modified wood. Part 1: wetting properties of acetylated, furfurylated, and thermally modified wood. Holzforschung 64(3):295–304

de Meijer M, Haemers S, Cobben W, Militz H (2000) Surface energy determinations of wood: comparison of methods and wood species. Langmuir 16(24):935–939

DIN 52617 (1987) Determination of the water absorption coefficient of construction materials. DIN German Institute for Standardization, Berlin

Donath S, Militz H, Mai C (2006) Creating water-repellent effects on wood by treatment with silanes. Holzforschung 60:40–46

Englund F, Bryne LE, Ernstsson M, Lausmaa J, Wålinder M (2009) Some aspects on the determination of surface chemical composition and wettability of modified wood. In: Proceedings of the 4th European conference on wood modification, Stockholm, pp 553–560

Ghosh S, Militz H, Mai C (2009) The efficacy of commercial silicones against blue stain and mould fungi in wood. Eur J Wood Prod 67:159–167

Ghosh S, Militz H, Mai C (2013) Modification of *Pinus sylvestris* L. wood with quat- and amino-silicones of different chain lengths. Holzforschung 67(4):421–427

Gindl M, Sinn G, Gindl W, Reiterer A, Tschegg S (2001) A comparison of different methods to calculate the surface free energy of wood using contact angle measurements. Colloids Surf A 181:279–287

Gsöls I, Raetzsch M, Ladner C (2003) Interactions between wood and melamine resins—effect on dimensional stability properties and fungal attack. In: Van Acker J, Hill C (eds) Proceedings first european conference on wood modification. Ghent, Belgium, pp 221–225

Hakkou M, Pétrissans M, El Bakali I, Gérardin P, Zoulalian A (2005) Wettability changes and mass loss during heat treatment of wood. Holzforschung 59:35–37

Hill CAS (2006) Wood modification: chemical, thermal and other processes. Wiley, Chichester

Hunt CG, Brandon R, Ibach RE, Frihart CR (2007) What does bonding to modified wood tell us about adhesion? In: Proceedings 5th COST E34 international workshop: bonding of modified wood. Bled, Slovenia, pp 47–56

Johansson J, Kifetew G (2010) CT-scanning and modelling of the capillary water uptake in aspen, oak and pine. Eur J Wood Prod 68(1):77–85

Kielmann BC, Adamopoulos S, Militz H, Koch G, Mai C (2014) Modification of three hardwoods with an N-methylol melamine compound and a metal-complex dye. Wood Sci Technol 48:123–136

Kocaefe D, Poncsak S, Doré G, Younsi R (2008) Effect of heat treatment on the wettability of white ash and soft maple by water. Holz Roh Werkst 66:355–361

Mahnert K-C, Adamopoulos S, Koch G, Militz H (2013) Topochemistry of heat-treated and N-methylol melamine-modified wood of koto (*Pterygota macrocarpa* K. Schum.) and limba (*Terminalia superba* Engl. et. Diels). Holzforschung 67(2):137–146

Mai C (2010) Processes of chemical wood modification **(in German)**. Holztechnologie 51:22–26

Mai C, Militz H (2004) Modification of wood with silicon compounds. Treatment systems based on organic silicon compounds—a review. Wood Sci Technol 37(6):453–461

Mantanis GI, Young RA (1997) Wetting of wood. Wood Sci Technol 31:339–353

Metsä-Kortelainen S, Antikainen T, Viitaniemi P (2006) The water absorption of sapwood and heartwood of Scots pine and Norway spruce heat-treated at 170 °C, 190 °C, 210 °C and 230 °C. Holz Roh Werkst 64:192–197

Militz H (2002) Thermal treatment of wood: European processes and their background. International Research Group on Wood Preservation, IRG/WP 02-40241

Neumann AW, Spelt JK (1996) Applied surface thermodynamics. Marcel Dekker Inc, New York

Niemz P (2007) Thermisch vergütetes Holz in der Schweiz. (Thermally treated wood in Switzerland) **(in German)**. Holz-Zentralblatt 133(40):1102

Nuopponen M, Vuorinen T, Jämsä S, Viitaniemi P (2003) The effect of heat treatment on the behaviour of extractives in softwood studied by FTIR spectroscopic methods. Wood Sci Technol 37:109–115

Nussbaum RM (1999) Natural surface inactivation of Scots pine and Norway spruce evaluated by contact angle measurements. Holz Roh Werkst 57:419–424

Owens DK, Wendt RC (1969) Estimation of the surface free energy of polymers. J Appl Polym Sci 13:1741–1747

Pétrissans M, Gérardin P, El bakali I, Serraj M (2003) Wettability of heat-treated wood. Holzforschung 57(3):301–307

Pries M, Wagner R, Kaesler K-H, Militz H, Mai C (2013) Acetylation of wood in combination with polysiloxanes to improve water-related and mechanical properties of wood. Wood Sci Technol 47:685–699

Rowell RM, Banks WB (1985) Water repellency and dimensional stability of wood. Gen Tech Rep FPL-50. Madison, WI

Scholz G, Krause A, Militz H (2009) Capillary water uptake and mechanical properties of wax soaked Scots pine. In: Proceedings of the 4th European conference on wood modification, Stockholm, pp 209–212

Sint KM (2010) Promoting utilization potential of *B. ceiba* Linn and *B. insigne* wall through enhancement of wood quality and technological properties by modification with melamine resin. Sierke Verlag, Thesis Univ. Göttingen

Sint KM, Adamopoulos S, Koch G, Hapla F, Militz H (2013) Impregnation of *Bombax ceiba* and *Bombax insigne* wood with a N-methylol melamine compound. Wood Sci Technol 47:43–58

Sivonen H, Maunu SL, Sundholm F, Jamsa S, Viitaniemi P (2002) Magnetic resonance studies of thermally modified wood. Holzforschung 56:648–654

Stamm AJ (1964) Wood and cellulose science. Ronald Press Company, New York

Treu A, Pilgård A, Puttmann S, Krause A, Westin M (2009). Material properties of furfurylated wood for window production. International Research Group on Wood protection, IRG/WP 09-40480

van Oss CJ, Giese RF Jr, Good RJ (1990) Reevaluation of the surface tension components and parameters of polyacetylene from contact angles of liquids. Langmuir 6(11):1711–1713

Wålinder MEP, Bryne LE (2006) Wood adhesion mechanisms: prediction of wood-thermoplastic-water interactions. In: CR Frihart (ed) Wood adhesives 2005. Forest Products Society. Proceedings No 7230, Madison WI, US, pp 385–392

Wålinder MEP, Johansson I (2001) Measurement of wood wettability by the Willhelmy method. Part 1. Contamination of probe liquids by extractives. Holzforschung 55:21–32

Weigenand O, Militz H, Tingaut P, Sèbe G, De Jeso B, Mai C (2007) Penetration of amino-silicone micro- and macro-emulsions into Scots pine sapwood and the effect on water-related properties. Holzforschung 61(1):51–59

Xiao Z, Xie Y, Militz H, Mai C (2010) Effect of glutaraldehyde on water related properties of solid wood. Holzforschung 64(4):483–488

Xie Y, Hill CAS, Xiao Z, Militz H, Mai C (2011) Dynamic water vapour sorption properties of wood treated with glutaraldehyde. Wood Sci Technol 45(1):49–61

Żenkiewicz M (2007) Methods for the calculation of surface free energy of solids. J Achiev Mater Manuf Eng 24:137–145

Zhang Y, Jin J, Wang S (2007) Effects of resin and wax on the water uptake behavior of wood strands. Wood Fiber Sci 39(2):271–278

Paper 3

Eur. J. Wood Prod. (2015) 73:635–642
DOI 10.1007/s00107-015-0920-2

ORIGINAL

Gross adhesive penetration in furfurylated, *N*-methylol melamine-modified and heat-treated wood examined by fluorescence microscopy

Alireza Bastani[1] · Stergios Adamopoulos[2] · Holger Militz[1]

Received: 1 August 2014 / Published online: 6 May 2015
© Springer-Verlag Berlin Heidelberg 2015

Abstract This study investigated the radial penetration of three conventional cold-set wood adhesives [emulsion polymer isocyanate (EPI), poly (vinyl acetate) (PVAc), one-component polyurethane (PU)] into various degrees of furfurylated and *N*-methylol melamine-modified (NMM) Scots pine, and heat-treated Scots pine and beech based on measurements of effective (EP) and maximum penetration (MP) from microscopic observations. EP of EPI adhesive decreased after modification with higher concentration of furfuryl alcohol while an improved penetration was recorded for PVAc into furfurylated wood. A deeper penetration was observed for all adhesives into wood treated with lower concentration of furfuryl alcohol. The EP of EPI and PU adhesives reduced after NMM treatment but it increased in the case of PVAc. In spite of reduction of EP of PU after NMM treatment, it represented a deeper penetration among all adhesives possibly due to its lower molecular weight. For Scots pine, increasing the treatment temperature improved EP of all adhesives while for beech, the EP of PU and PVAc increased largely in the case of samples treated at 195 °C. Visual analysis of fluorescence microscopy pictures provided more detailed information on modality of penetration. The results are useful for understanding the interaction among common adhesives and modified materials, and can be used in future research to explain the bonding behavior of modified wood.

✉ Alireza Bastani
abastan@gwdg.de

[1] Wood Biology and Wood Products, Burckhardt Institute,
Georg-August-University Göttingen, Büsgenweg 4,
37077 Göttingen, Germany

[2] Department of Forestry and Wood Technology, Linnaeus
University, 351 95 Växjo, Sweden

1 Introduction

Penetration of adhesives into the wood structure plays an important role in the production of glued wood-based panels and products by affecting the bond quality (Frihart 2005; Kamke and Lee 2007). An appropriate amount of adhesive penetration is always desired for producing effective bonding of wood particles and layers by providing the necessary surface contact for mechanical interlocking, covalent bonding or secondary interactions (Pizzi 1994). Insufficient penetration of adhesive causes minimal surface contact leaving a thick film of adhesive on the surface whereas over penetration creates starved bondlines, thus leading to poor bonding in both cases (Johnson and Kamke 1992).

As defined by Sernek et al. (1999), adhesive penetration is the spatial distance from the wood surface to maximum observable depth into the wood. It can be categorized as gross penetration (micrometer scale) and cell wall infiltration (nanometer scale). Gross penetration is defined as the movement of liquid adhesive into the porous structure of wood, mainly filling lumens, large voids and pits, and is affected by hydrodynamic flow and capillary suction of wood network tissue. The main anatomical wood elements having a sufficient lumen space for conducting liquid adhesive flow are vessels for hardwoods and longitudinal tracheids for softwoods. Cell wall infiltration consists of movement of glue into tiny voids and microstructure of the wood cell wall (Fengel and Kumar 1970; Kamke and Lee 2007). Adhesive penetration within the porous wood structure, fractures in wood surface cells, and cell walls leads to the development of large regions of mechanical interlocking at macro-, micro- and nano scales, thus enhancing the mechanical bond strength at or near the interface (Frihart 2006). According to Marra (1992), adhesive penetration can be influenced by wood related factors (e.g. porosity, direction of penetration, wood species, earlywood and latewood, sapwood and heartwood, moisture content, surface energy), adhesive related factors (e.g. type, viscosity, formulation), and process related factors (e.g. amount and duration of application of pressure, temperature, time of open assembly). The ultimate bonding performance of the wood bond is the result of interaction among these factors.

Several investigations have been carried out over the years to evaluate adhesive penetration into the macro- and micro-structure of wood by using various methods, mainly microscopy of cross-sections and micro-slides (Johnson and Kamke 1992; Sernek et al. 1999). Scanning electron microscopy (SEM) has also been widely used to examine the adhesive penetration into microscopic cell cavities (Koran and Vasishth 1972; Saiki 1984), sometimes supported by energy-dispersive X-ray analysis (Smith and Côté 1971; Bolton et al. 1988). While it is possible to have a three-dimensional view of wood tissues by using SEM microscopy, the problem is to visually distinguish between wood and adhesive (Modzel et al. 2011). Other more sophisticated methods were performed using porosimetry (Wang and Yan 2005), neutron radiography (Niemz et al. 2004), and synchrotron radiation X-ray tomographic microscopy (Hass et al. 2012). Detection of adhesive penetration into the cell wall has used scanning thermal microscopy (Konnerth et al. 2008), UV-microscopy (Gindl et al. 2002), energy loss spectroscopy (Rapp et al. 1999), and confocal laser scanning microscopy (Xing et al. 2005).

Basic knowledge of the bonding behavior of modified wood has already been established, despite the complexity posed by the great variety of adhesives, species, and modification methods. The adhesive strength might be affected negatively by the poorer adhesive wetting of the less polar and less porous modified wood surfaces (except the more porous structure of heat-treated wood) as well as by the fewer chemical bonds between the two surfaces (Hunt et al. 2007). For example, acetylation of wood results in the loss of hydroxyl groups, making the wood more hydrophobic and thus reduces the hydrogen bonding with the adhesive (Brandon et al. 2005). Chemical modification of wood components, especially hemicelluloses and lignin, occurring during heat treatment reduce the hygroscopicity of wood and consequently alter the distribution and penetration of the adhesive in wood. Depending on the adhesive type used, effects on the curing process of adhesives might occur due to the decreased pH and wettability of heat-treated wood (Boonstra et al. 1998; Sernek et al. 2008; Sahin Kol et al. 2009; Boruszewski et al. 2011). The presence of silicone compounds can delay or prevent the adhesive wetting of silicone modified wood surfaces and proper curing thereafter (Kurt et al. 2008). Impregnation

modification with different resin systems, like for example phenol–formaldehyde resin, providing deposition of resins into the cell wall and lumens as well as hydrophobic effects, can also affect bonding (Adamopoulos et al. 2012). Positive effects on bonding strength might stem from the reduced shrinking and swelling stresses of modified wood on the cured adhesive bond (Frihart 2006; Atar et al. 2007; Sernek et al. 2008). It has been shown that the adhesive type plays a primary role in a satisfactory gluing of modified wood. This fact has been reported for a number of cases, such as for acetylated (Vick and Rowell 1990; Vick et al. 1993; Frihart et al. 2004), furfurylated (van der Zee et al. 2007), phenol–formaldehyde modified (Adamopoulos et al. 2012), silicone modified (Kurt et al. 2008), and heat-treated wood (Boonstra et al. 1998; Sernek et al. 2008; Sahin Kol et al. 2009). The interaction of negative and positive effects of the different modifications mentioned previously and the many different adhesives used increase the complexity of drawing definite conclusions from the bonding performance of modified wood. Thus, a more in-depth study of the gluing properties of modified wood is needed to reveal why specific adhesive systems provide adequate bonds while others fail, and possibly to lead to adaptations of the bonding process. For example, Adamopoulos et al. (2012) stated that the amount of adhesive penetration into the modified wood substrate has been little studied as an influencing parameter. For the adhesives phenol–resorcinol–formaldehyde and poly (vinyl acetate) (PVAc), the penetration into the porous wood structure was the lowest at highest degree of phenol–formaldehyde modification. This resulted in significant lower adhesion only for PVAc. Penetration of various adhesives was adequate in acetylated wood, but flow into cell lumens did not always relate to bond strength (Chandler et al. 2005).

Wood modification can improve dimensional stability and resistance to biological degradation and moisture, but can also create a new surface for bonding. With increasing use of modified wood for exterior and interior applications, it is not only important to measure the strength of modified wood bonded with various adhesives but also to know the interaction of adhesive polymers with wood tissues. Although application of pressure is important in the bonding process for curing of some adhesives, as a first step it is equally important to understand the interaction between modified wood and adhesive. Specifically, it is necessary to establish baseline knowledge of the ability of various adhesives to penetrate into modified wood without application of pressure. Therefore, the aim of this study was to detect adhesive penetration into different modified wood materials by using fluorescence microscopy. Scots pine and beech as common wood species used in furniture and construction applications, furfurylation, melamine and heat treatment being among the commercial wood modifications, and three main solid wood adhesives (two water-borne, one moisture curing) used widely in wood-working industry were selected for this study. This combination of wood species, modification methods and adhesives is expected to provide sufficient information on the interaction of adhesives and modified wood.

2 Materials and methods

2.1 Wood material and modification methods

The modified wood materials used were furfurylated Scots pine (*Pinus sylvestris* L.), melamine-treated Scots pine, and heat-treated Scots pine and beech (*Fagus sylvatica* L.). Furfurylation was done by Kebony ASA (Skien, Norway) following an industrial process with two concentrations of furfuryl alcohol, which are industrially known as Kebony FA 40 and FA 70, with a 65 and 75 % weight percent gain (WPG), respectively. For the melamine modification, the *N*-methylol melamine (NMM) resin Madurit MW840/75WA (Ineos Melamines GmbH, Frankfurt, Germany) was provided as an aqueous stock solution with a solid content of approx. 75 %. Solutions of 10, 20 and 30 % NMM solid content were prepared and impregnation of wood was done in a stainless steel vessel using a full cell process, which included an initial vacuum phase of 100 mbar for 1 h, a pressure phase of 12 bar for 2 h, and finally curing of the NMM resin at 120 °C for 1 week in a drying chamber. Heat-treatment was carried out with the industrial scale vacuum-press dewatering method (Vacu³) at 195 and 210 °C by Timura Holzmanufaktur GmbH (Südharz, Germany).

The modified wood materials were conditioned at 20 °C and 65 % RH in a climate chamber, and then planned samples were prepared with dimensions of 15 × 30 × 60 mm³ (radial × tangential × longitudinal).

2.2 Section preparation and fluorescence microscopy

Emulsion polymer isocyanate (EPI), one-component polyurethane (PU) and poly (vinyl acetate) (PVAc) (Table 1, all provided by Jowat AG, Detmold, Germany) were spread uniformly on the tangential surfaces of wood samples at a loading level of 200 g/m² by hand brushing. The samples were left aside for 1 week at room temperature (20 °C), thus leaving adhesives to penetrate as much as possible. Although it is known that most of the penetration normally happens within a few minutes after deposition of the adhesive on the wood surface even before its hardening (Sernek et al. 1999; Kamke and Lee 2007), a long exposure time of 1 week was selected as to provide

Table 1 Information on adhesives

Property	Adhesive		
	EPI	PU	PVAc
Solids, (%)	60	99.5	49
Brookfield viscosity (20 °C), (MPa)	9400[a]	10,500	5000[a]
Density, (g/cm³)	1.5	1.15	1.04
pH	7	–	5.2

[a] After adding cross linking agent (hardener)

the necessary time needed for all the different adhesive types used (the two waterborne PVAC and EPI, and the moisture curing PU), which might react and penetrate wood differently, to reach the maximum possible penetration. In addition, a long contact time was preferable as drying of the adhesives on the wood surface would not allow judging visually whether the penetration had stopped or not.

To quantify the radial adhesive penetration (lumen filling) into the porous wood structure, 40-µm-thick sections were cut on a Reichert-Jung sliding microtome exposing a glueline with a transverse surface within earlywood or latewood for Scots pine and in random positions within the growth ring for beech. After staining with a pipette drop of 0.5 % safranin O solution, the sections were placed on glass slides and examined under an Eclipse 50i fluorescence microscope with appropriate filter sets, equipped with a Sight DS-5 M-L1 digital camera and a NIS-Elements F software for image analysis (Nikon, Düsseldorf, Germany).

Random areas of penetrated adhesive were used to measure effective penetration (EP) and maximum penetration (MP) following the definitions of Sernek et al. (1999). According to this method, EP is the whole area of adhesive detected in the interphase region of the bondline divided by the width of the glueline. Cell walls and unfilled lumen areas are excluded from the calculation of EP. MP is the mean distance of penetration of the five most remote adhesive objects found inside the field of view. NIS-elements F software provides the possibility to separate the bright adhesive objects from the darker background and to calculate the required statistical parameters. For each case, ten regions of interest with dimensions of 1100 × 600 µm² (tangential × radial) were selected to measure EP and MP using the formulas (see also Fig. 1):

$$EP = \frac{\sum_i^n Ai}{Xo} \tag{1}$$

where EP is the effective penetration, µm; Ai is the area of adhesive object i, µm²; n is the number of objects; Xo is the width of the maximum rectangle determining the measurement area (1100 µm),

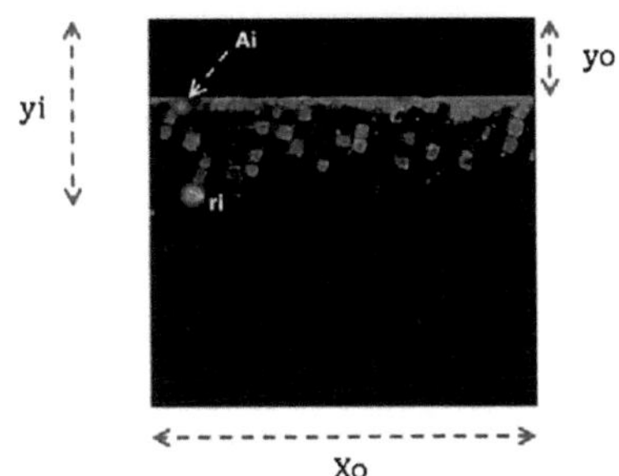

Fig. 1 Illustration of EP and MP components of gross adhesive penetration (radial) on a transverse section

$$MP = \frac{\sum_i^5 (yi + ri - yo)}{5} \qquad (2)$$

where MP = maximum depth of penetration, μm; yi = centroid of adhesive object i representing the deepest penetration, μm; ri = mean radius of adhesive object i, μm; yo = reference y-coordinate of the glueline interface, μm. Each obtained EP and MP value represented the average of 50 measurements and the selected regions of interest could cover a large area in micron level scale.

The selected method is known for determining adhesive flow into wood lumens with or without applying pressure (e.g. Johnson and Kamke 1992; Sernek et al. 1999). In this case, it is expected to provide valuable information on the interaction between the adhesives and the hydrophobic modified surfaces and substrates, especially for EPI and PVAc adhesives which mostly cure physically by losing water. In order to gain a better understanding of adhesive flow, the penetration and localisation of adhesives into the wood structure was analyzed based on visual observation of corresponding photomicrographs for each case.

3 Results and discussion

3.1 Adhesive penetration in the radial direction

As shown in Table 2, furfurylation was not found to affect the effective penetration (EP) of PU adhesive. The same was also true for the EP of EPI adhesive but only for the treatment with lower uptake of furfuryl alcohol (FA 40). Furfurylation at higher WPG (FA 70) almost halved EP of EPI as compared to control, while both types of furfuryl alcohol increased significantly EP of PVAc adhesive into the wood structure. A possible explanation for the increase of EP of the water-based and low viscosity PVAc in furfurylated wood could be the presence of hygroscopic

buffering salt agents used in the furfurylation process (Bryne and Wålinder 2010), which might have caused an increased wetting of this type of adhesive. Another explanation could be the lower rate of water uptake by furfurylated wood (Treu et al. 2009), which causes the PVAc glue to remain more liquid for a longer time on the wood surface and hence providing more time to flow into the wood tissue. However, this trend was not confirmed by the other waterborne adhesive used (EPI). For all types of adhesives, maximum penetration (MP) was slightly higher in the FA 40 treated wood than in the FA 70 (Fig. 2). In a similar mode of action, it has previously been reported that a higher WPG of furfuryl alcohol causes a slightly greater reduction in water uptake (Treu et al. 2009).

The EP of PU adhesive was reduced significantly and equally by every NMM modification at a more or less same level (Table 2). The EP of EPI was reduced only by the most intense NMM modification (30 %) and showed a very low value. A reduction in water uptake (total, both lumens and cell-walls) in NMM-modified wood has previously been reported, which did not follow a certain pattern based on the concentration of the NMM treatment solution (Sint 2010). In contrast, PVAc showed a significantly higher EP by the 10 and 20 % NMM treatments and it remained unchanged in the 30 % NMM-treated wood. This behaviour of the water-based PVAc adhesive is difficult to explain, as it would be expected that the incorporation of NMM resin into the wood structure would impair adhesive wetting. It seems that in NMM modified wood it is still possible for PVAc to flow through different morphological regions of wood tissues as NMM resin results mainly in cell wall bulking rather than occlusion of cell lumens (Lukowsky 1999; Kielmann et al. 2014), which are the main penetration pathways for liquids. The highest MP was noted for PU, especially for 20 and 30 % NMM, although this adhesive showed a reduced EP by the NMM modification (Fig. 2).

Heat-treatment generally increased significantly the EP of adhesives in Scots pine (Table 2). EP was approximately doubled as compared to untreated wood for PU adhesive at the lower treatment temperature of 195 °C, and for EPI at 210 °C, while it was increased approximately four to five times for PU and PVAc adhesives in wood heat-treated at 210 °C. An exception was noted for the significant reduction in EP of EPI in Scots pine heat-treated at 195 °C as well as for the unchanged EP of PVAc at the same temperature. The situation was a bit different in the case of heat-treated beech (Table 2). For EPI adhesive, EP was significantly reduced to more than half of the untreated value for both treatment temperatures. A similar EP behaviour was observed for PU and PVAc adhesives. EP was significantly increased for beech treated at 195 °C but did not change at 210 °C. It should be expected a decrease in

Table 2 Effective penetration (EP) of EPI, PU and PVAc adhesives in the radial direction of furfurylated Scots pine (FA 40 and 70), NMM modified Scots pine (10, 20 and 30 % NMM), and heat-treated Scots pine (HT-S) and beech (HT-B)

Treatment	EP, µm		
	EPI	PU	PVAc
Scots pine			
Scots pine-control	9.37 (3.82)a	9.75 (7.35)a	3.95 (3.39)a
FA 40	11.64 (3.25)a	7.06 (1.81)a	11.14 (7.08)b
FA 70	5.07 (2.80)b	7.13 (4.18)a	9.58 (3.49)b
p value	0.001	0.402	0.008
10 % NMM	7.20 (6.36)ab	4.85 (2.34)b	9.90 (5.63)b
20 % NMM	5.96 (4.16)ab	5.92 (2.29)b	11.31 (4.43)b
30 % NMM	2.40 (2.00)b	4.89 (1.21)b	4.52 (2.92)a
p value	0.009	0.033	0.000
HT-S 195 °C	1.42 (0.63)b	24.79 (5.50)b	3.22 (1.24)a
HT-S 210 °C	18.10 (2.70)c	43.37 (13.91)c	20.29 (10.30)b
p value	0.000	0.000	0.000
Beech			
Beech-control	25.06 (18.94)a	17.46 (3.91)a	13.90 (3.90)a
HT-B 195 °C	10.35 (6.73)b	48.73 (13.62)b	31.01 (14.25)b
HT-B 210 °C	11.19 (3.55)b	23.07 (8.29)a	16.61 (9.61)a
p value	0.015	0.000	0.000

Mean values and standard deviations in parenthesis. Values followed by a different letter within a column are statistically different (ANOVA and Tukey HSD test)

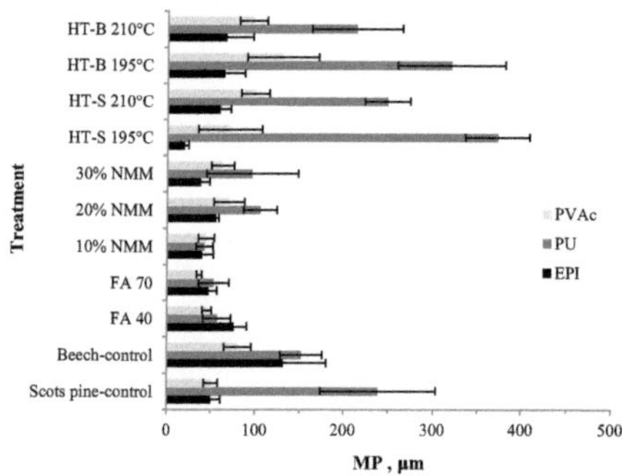

Fig. 2 Maximum penetration (MP) of EPI, PU and PVAc adhesives in the radial direction of furfurylated Scots pine (FA 40 and 70), NMM modified Scots pine (10, 20 and 30 % NMM), and heat-treated Scots pine (HT-S) and beech (HT-B). *Error bars* represent 95 % confidence intervals

adhesive wetting of heat-treated wood as a result of thermal degradation of wood polymers, mainly hemicelluloses (Militz 2002; Hakkou et al. 2005). It is quite difficult to explain the increase in EP of adhesives as presented above. Wood defects that might have occurred during the heat treatment, such as cracking and broken cell walls (Boonstra et al. 2006a, b), could have contributed to this result by providing flow pathways at various spots. However,

fluorescence microscopy presents limitations in detecting tiny cracks at the cell walls. MP of adhesives did not follow a certain pattern with the treatment temperatures (Fig. 2). In heat-treated Scots pine, MP values of EPI and PVAc adhesives were increased at the high treatment temperature of 210 °C and reduced in the case of PU adhesive. For beech, MP value of PU and PVAc were found to increase while it remained stable for EPI.

3.2 Microscopic observations

Microscopic observation of wood-adhesive interfaces provided additional information on the radial penetration of adhesives in various wood tissues and cell types. It should be acknowledged that while optical fluorescence microscopy can easily ascertain wood adhesive penetration into microscopic cell cavities, it has limitations in capturing the complex three-dimensional wood structure, and thus in providing full details of adhesive flow. It should also be taken into account that the adhesives used are pre-polymerized adhesives, and they are mainly consisting of higher molecular weight molecules. Thus, they can enter the cell lumens rather than infiltrating into the cell wall.

EPI adhesive, having a high viscosity, could penetrate radially more easily in earlywood by forming continuous layers of filled cells (up to 2–3 cells in depth) than in latewood zones of untreated Scots pine, obviously through the larger pits of axial tracheids and ray cells. For untreated

beech, EPI could hardly penetrate the narrow latewood zones at ring borders with thicker fibre cell walls and small-diameter vessels. This penetration behaviour of EPI was also observed in furfurylated Scots pine for both treatments, FA 40 (Fig. 3a) and FA 70, as well as for 10 and 20 % NMM-treated Scots pine. In 30 % NMM-treated

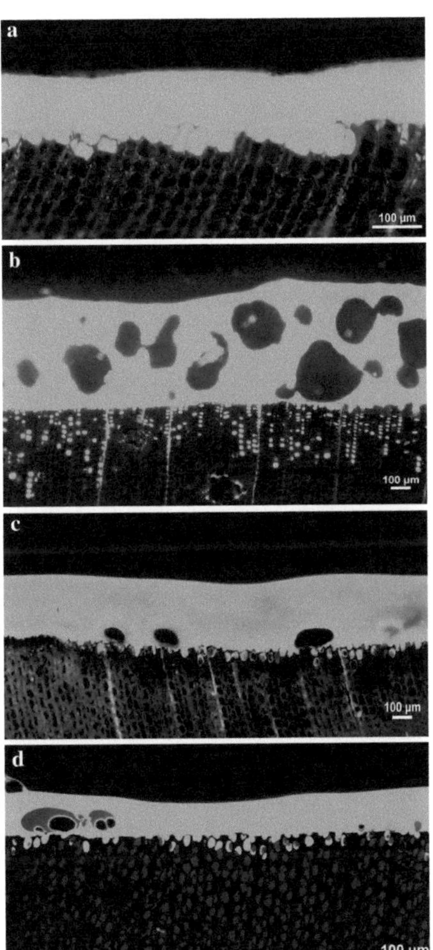

Fig. 3 Cured adhesives on transverse surfaces of modified wood. EPI and furfurylated (FA 40) Scots pine (a); PU and heat-treated Scots pine at 195 °C (b); PVAc and 20 % NMM Scots pine (c); PVAc and heat-treated beech at 195 °C (d). *Scale bars* = 100 µm

wood, EPI could hardly penetrate even the earlywood zones probably because the resin at this high concentration in the impregnation solution might have filled all the voids in the cell wall and might also have formed deposits in the lumens due to self-condensation of the excess NMM (Lukowsky 1999). It is not clear at what point the added NMM exceed the cell wall absorption volume and start to accumulate and occlude the lumens, but the phenomenon has been depicted previously in hardwood cell lumens by means of light microscopy and SEM and confirmed by cellular UV microspectrophotometry (Mahnert et al. 2013; Kielmann et al. 2014). Heat-treated Scots pine at 195 °C showed almost no EPI penetration in both earlywood and latewood zones. On the contrary, EPI penetration of heat-treated Scots pine at 210 °C was characterized by a continuous layer with fully penetrated cells similar to that of the untreated Scots pine. Concerning heat-treated beech, penetration of EPI was restricted to scattered vessels filled with the adhesive and only very few filled parenchyma cells close to the glueline.

PU adhesive, having a lower molecular weight than EPI, could penetrate earlywood and latewood zones of untreated Scots pine by forming layers made up of a few filled or partly filled tracheids and ray cells in depth. The molecular weight of the adhesive might play a role in impregnation modification of wood as the used chemicals might also be deposited and partly occlude the cell lumens (Kielmann et al. 2014). A similar penetration mode was observed for untreated beech with filled and partly filled vessels. In the case of furfurylation and NMM treatments, under the microscope PU penetration looked similar to that of the untreated wood with only a few filled or partially filled cells with maximum up to 1–2 cells in depth. A relatively deep penetration zone (up to 7–20 cells) in both earlywood and latewood with fully and partly filled tracheids and ray cells in scattered radial lines characterized the radial flow of PU in heat-treated Scots pine at both temperatures, 195 (Fig. 3b) and 210 °C. Deep PU penetration layers with filled and partly filled adjacent vessels and axial parenchyma cells were also found in beech heat-treated at 195 °C. Less but similar PU penetration was noted for the higher treatment temperature of 210 °C. It seems that there were less functional –OH groups in the heat-treated wood available for chemical bonding (Hill 2006; Nguila Inari et al. 2007; Li et al. 2011) with the reactive PU adhesive than in the melamine and furfurylated wood, leading to higher penetration in heat-treated wood as a result of less reaction on the surface.

Observing the penetration of PVAc adhesive into untreated, furfurylated, and NMM-treated Scots pine wood, it can be stated that the penetration zone in both earlywood and latewood consisted of a continuous uneven layer with filled or partially filled cells with mostly one and few positions with two cells in depth (Fig. 3c). Heat-treated Scots pine and beech wood had the same PVAc penetration pattern as mentioned before. A considerably higher penetration was recorded for Scots pine for treatment at 210 °C and for beech at 195 °C (Fig. 3d), where at positions the penetrated layer of adhesive could reach up to three cells in depth.

The overall results provided an understanding of the interaction of common adhesives and modified materials. It is apparent that these findings cannot be related directly and explain the bonding behaviour of modified wood. The specific radial penetration data and modality of penetration within different morphological regions of wood tissues for each adhesive type and modified material could provide useful explanations about the bonding strength in future studies and its correlation with parameters of adhesive penetration induced by pressure. It should also be remembered that adhesive penetration is one of the parameters affecting the creation of a strong bond (Frihart 2005). However, there is a general belief that by increasing adhesive penetration into the wood structure the ultimate bond quality would increase (Johnson and Kamke 1992; Kamke and Lee 2007).

4 Conclusion

Based on the results of gross radial penetration of EPI, PU and PVAc adhesives into different modified wood materials as well as on microscopic observations of the adhesive flow into the coarse capillary structure of wood, the following conclusions can be drawn:

- Furfurylation had a negative effect only on EP of EPI into FA 70 treated Scots pine wood. In contrary, the EP of PVAc was elevated after modification with both FA 40 and FA 70 solutions. All adhesives showed a deeper penetration into wood treated with lower concentration of furfuryl alcohol.
- The EP of PU reduced after NMM modification and also that of EPI but only after treatment with the highest concentration of NMM (30 %). Like furfurylation, the EP of PVAc increased after 10 and 20 % NMM treatment. Despite the reduction in EP of PU after NMM modification, this adhesive penetrated deeper than EPI and PVAc into NMM-treated wood due to its lower molecular weight.
- After heat-treatment of Scots pine wood, the EP of all adhesives increased considerably with the exception of EPI and PVAc into samples treated at 195 °C. In the case of heat-treated beech, the EP of EPI was decreased but it was increased considerably for PU and PVAc at the lower treatment temperature of 195 °C. PU penetrated radially much deeper into heat-treated Scots

pine and beech for both treatment temperatures than the two other adhesives.

The quantitative data of EP and MP obtained in this study provided valuable information on the interaction among common adhesives and modified materials, and can be useful in revealing the parameters affecting the bonding strength of modified wood in future research.

References

Adamopoulos S, Bastani A, Gascon-Garrido P, Militz H, Mai C (2012) Adhesive bonding of beech wood modified with a phenol formaldehyde compound. Eur J Wood Prod 70:897–901

Atar M, Keskin H, Yavuzcan HGM (2007) Impact of impregnation with boron compounds on the bonding strength of some wood materials. Proceedings 5th COST E34 International Workshop: bonding of modified wood. Bled, Slovenia, pp 61–69

Bolton AJ, Dinwoodie JM, Davies DA (1988) The validity of the use of SEM/EDAX as a tool for the detection of UF resin penetration into wood-cell walls in particleboard. Wood Sci Technol 22:345–356

Boonstra MJ, Tjeerdsma BF, Groeneveld HAC (1998) Thermal modification of non-durable wood species. Part 1, the Plato technology: thermal modification of wood. International research group on wood preservation, document no. IRG/WP 98-40123

Boonstra MJ, Rijsdijk JF, Sander C, Kegel E, Tjeerdsma B, Militz H, van Acker J, Stevens M (2006a) Microstructural and physical aspects of heat treated wood. Part 1. Softwoods Maderas Cienc Tecnol 8:193–208

Boonstra MJ, Rijsdijk JF, Sander C, Kegel E, Tjeerdsma B, Militz H, van Acker J, Stevens M (2006b) Microstructural and physical aspects of heat treated wood. Part 2. Hardwoods Maderas Cienc Tecnol 8:209–217

Boruszewski PJ, Borysiuk P, Mamiński MŁ, Grześkiewicz M (2011) Gluability of thermally modified beech (*Fagus silvatica* L.) and birch (*Betula pubescens Ehrh.*). Wood Mater Sci Eng 6(4):185–189

Brandon R, Ibach RE, Frihart CR (2005) Effects of chemically modified wood on bond durability. In: Frihart CR (ed) Wood adhesives 2005. Forest Products Research Society, San Diego, pp 111–114

Bryne LE, Wålinder MEP (2010) Ageing of modified wood. Part 1: wetting properties of acetylated, furfurylated, and thermally modified wood. Holzforschung 64:295–304

Chandler JG, Brandon LR, Frihart CR (2005) Examination of adhesive penetration in modified wood using fluorescence microscopy. ASC spring 2005 convention and exposition: April 17–20. Columbus, p 10

Fengel D, Kumar RN (1970) Electron microscopic studies of glued wood joints. Holzforschung 24:177–181

Frihart CR (2005) Wood adhesion and adhesives. In: Rowell RM (ed) Handbook of wood chemistry and wood composites. CRC Press LLC, Boca Raton, pp 215 278

Frihart CR (2006) Wood structure and adhesive bond strength. In: Stokke DD, Groom LH (eds) Characterization of the cellulosic cell wall. Blackwell Publishing, Ames, pp 241–253

Frihart CR, Brandon R, Ibach RE (2004) Selectivity of bonding for modified wood. P Adhes Soc 27(1):329–331

Gindl W, Dessipri E, Wimmer R (2002) Using UV-microscopy to study diffusion of melamine-urea-formaldehyde in cell walls of spruce wood. Holzforschung 56:103–107

Hakkou M, Pétrissans M, El Bakali I, Gérardin P, Zoulalian A (2005) Wettability changes and mass loss during heat treatment of wood. Holzforschung 59:35–37

Hass P, Wittel FK, Mendoza M, Herrmann HJ, Niemz P (2012) Adhesive penetration in beech wood: experiments. Wood Sci Technol 46(1–3):243–256

Hill CAS (2006) Wood modification: chemical, thermal and other processes. Wiley, Chichester

Hunt CG, Brandon R, Ibach RE, Frihart CR (2007) What does bonding to modified wood tell us about adhesion? Proceedings 5th COST E34 International Workshop: bonding of modified wood. Bled, Slovenia, pp 47–56

Johnson SE, Kamke FA (1992) Quantitative analysis of gross adhesive penetration in wood using fluorescence microscopy. J Adhes 40:47–61

Kamke FA, Lee JN (2007) Adhesive penetration of wood—a review. Wood Fiber Sci 39(2):205–220

Kielmann BC, Adamopoulos S, Militz H, Koch G, Mai C (2014) Modification of three hardwoods with an *N*-methylol melamine compound and a metal-complex dye. Wood Sci Technol 48:123–136

Konnerth J, Harper D, Lee SH, Rails TG, Gindl W (2008) Adhesive penetration of wood cell walls investigated by scanning thermal microscopy (SThM). Holzforschung 62:91–98

Koran Z, Vasishth RC (1972) Scanning electron microscopy of plywood glue lines. Wood Fiber 3(4):202–209

Kurt R, Mai C, Krause A, Militz H (2008) Hydroxymethylated resorcinol (HMR) priming agent for improved bondability of silicone modified wood glued with a polyvinyl acetate adhesive. Holz Roh Werkst 66:305–307

Li X, Cai Z, Mou Q, Wu Y, Yuan Liu (2011) Effects of heat treatment on some physical properties of Douglas fir (Pseudotsuga menziesii) wood. Adv Mater Res 197–198:90–95

Lukowsky D (1999) Holzschutz mit Melaminharzen (wood protection with melamine resins). PhD thesis, University of Hamburg

Mahnert K-C, Adamopoulos S, Koch G, Militz H (2013) Topochemistry of heat-treated and *N*-methylol melamine modified wood of Koto (*Pterygota macrocarpa* K. Schum.) and Limba (*Terminalia superba* Engl. et Diels). Holzforschung 67(2):137–146

Marra AA (1992) Technology of wood bonding. Van Nostrand Reinhold, New York, pp 35–54

Militz, H. (2002) Thermal treatment of wood: European processes and their background. International research group on wood preservation, IRG/WP 02-40241

Modzel G, Kamke FA, De Carlo F (2011) Comparative analysis of a wood-adhesive bondline. Wood Sci Technol 45:147–158

Nguila Inari G, Petrissans M, Gerardin P (2007) Chemical reactivity of heat-treated wood. Wood Sci Technol 41(2):157–168

Niemz P, Mannes D, Lehmann E, Vontobel P, Haase S (2004) Experiments on the distribution of adhesive close to the glue joint by neutron radiography and microscopy (in German). Holz Roh Werkst 62:424–432

Pizzi A (1994) Advanced wood adhesives technology. Marcel Dekker Inc., New York, p 289

Rapp AO, Bestgen H, Adam W, Peck RD (1999) Electron energy loss spectroscopy (EELS) for quantification of cell-wall penetration of a melamine resin. Holzforschung 53:111–117

Sahin Kol H, Özbay G, Altun S (2009) Shear strength of heat-treated tali (*Erythrophleum ivorense*) and iroko (*Chlorophora excelsa*) woods, bonded with various adhesives. Biores 4(4):1545–1554

Saiki H (1984) The effect of the penetration of adhesives into cell walls on the failure of wood bonding. Mokuzai Gakkaishi 30(1):88–92

Sernek M, Resnik J, Kamke FA (1999) Penetration of liquid urea-formaldehyde adhesive into beech wood. Wood Fiber Sci 31(1):41–48

Sernek M, Boonstra M, Pizzi A, Despres A, Gérardin P (2008) Bonding performance of heat treated wood with structural adhesives. Holz Roh Werkst 66:173–180

Sint KM (2010) Promoting utilization potential of *B. ceiba* Linn *and B. insigne* Wall through enhancement of wood quality and technological properties by modification with melamine resin. Sierke Verlag, thesis Univ. Göttingen

Smith LA, Côté WA (1971) Studies of penetration of phenol-formaldehyde resin into wood cell walls with the SEM and energy-dispersive X-ray analyser. Wood Fiber 3:56–58

Treu A, Pilgård, Puttmann S, Krause A, Westin M (2009) Material properties of furfurylated wood for window production. International research group on wood preservation, IRG/WP 09-40480

van der Zee ME, Schipholt NL, Tjeerdsma BF, Brynildesn P, Mohoric I (2007) Glueability and paintability of furfurylated wood (Kebony). In: Hill CAS (ed) Proceedings third european conference on wood modification. Cardiff, pp 231–234

Vick CB, Rowell RM (1990) Adhesive bonding of acetylated wood. Int J Adhes Adhes 10(4):263–272

Vick CB, Larsson PC, Mahlberg RL, Simonson R, Rowell RM (1993) Structural bonding of acetylated Scandinavian softwoods for exterior lumber laminates. Int J Adhes Adhes 13(3):139–149

Wang WQ, Yan N (2005) Characterising liquid resin penetration in wood using a mercury intrusion porosimeter. Wood Fiber Sci 37:505–514

Xing C, Riedl B, Cloutier A, Shaler SM (2005) Characterisation of urea-formaldehyde resin penetration into medium density fibreboard fibers. Wood Sci Technol 39:374–384

Paper 4

Contents lists available at ScienceDirect

International Journal of Adhesion & Adhesives

journal homepage: www.elsevier.com/locate/ijadhadh

Study of adhesive bondlines in modified wood with fluorescence microscopy and X-ray micro-computed tomography

Alireza Bastani [a,*], Stergios Adamopoulos [b], Tim Koddenberg [a], Holger Militz [a]

[a] Wood Biology and Wood Products, Burckhardt Institute, Georg-August-University Göttingen, Büsgenweg 4, 37077 Göttingen, Germany
[b] Department of Forestry and Wood Technology, Linnaeus University, Lückligs plats 1, 351 95 Växjo, Sweden

ARTICLE INFO

Article history:
Accepted 6 April 2016
Available online 13 April 2016

Keywords:
Modified wood
Adhesive penetration
Bondline
Fluorescence microscopy
X-ray tomography

ABSTRACT

The quantitative penetration of three coldset wood adhesives [one-component polyurethane (PU), emulsion polymer isocyanate (EPI), poly (vinyl acetate) (PVAc)] under hydraulic pressure into different types of modified wood was studied using fluorescence microscopy and the results were compared to these of a previous study without pressure on adjacent wood samples. The effective penetration (EP) of PU was negatively affected by furfurlylation and NMM modification when pressure was applied. For PVAc, 30% NMM treatment and heat treatment of Scots pine and beech at 210 °C had a negative effect on its EP, but against this the EP of this adhesive increased after heat treatment of beech at 195 °C. In the case of furfurylation, the depth of penetration of all adhesives was less into wood treated with higher concentration of furfuryl alcohol. PU showed a much deeper penetration into NMM-modified and heat-treated wood than the other adhesives with the exception of heat-treated beech at 195 °C. Application of pressure led to rather different results as compared to the EP data when no pressure was applied. The three-dimensional (3D) visualisation of the penetration of PU adhesive into heat-treated Scots pine was also examined by X-ray micro-computed tomography (XμCT). The 3D flow pattern of PU adhesive into heat-treated Scots pine was clearly depicted by XμCT.

© 2016 Elsevier Ltd. All rights reserved.

1. Introduction

Glued wood products have an important share in furniture and construction markets. Among the different factors involved in the wood bonding process, penetration of adhesives into the porous structure of wood is essential since it might influence the bond quality and subsequently the performance of the whole structure. An optimal amount of penetrated adhesive is essential also for economic reasons as the equivalent weight of adhesive is much more expensive than wood [21].

Adhesive penetration is defined as the movement of a fluid glue from the surface into the voids and porous structure of wood tissue, and it can be grouped as gross penetration (mainly filling cell lumens and large voids) and cell wall penetration (into the tiny voids and microstructure of wood cell wall).There are several mechanisms involved in adhesion such as mechanical interlocking, covalent bonding and secondary interactions (e.g. hydrogen bonds), which can be influenced by adhesive penetration [28]. Adhesive penetration also stabilizes the wood cells which are weakened due to surfacing and pressing. Excessive amount of

penetration can cause so-called "starved bondlines" that bring an imperfect or a very thin adhesive layer between the two adherents, while inadequate quantity of penetrated adhesive results in formation of a thick bondline and a weak link between this bondline and the internal area of wood [19]. Penetration of adhesives into the porous wood structure and cell walls has been widely studied within last years by using a variety of methods, such as optical microscopy [1,36,7], scanning electron microscopy [24,39,8], scanning thermal microscopy [23], UV-microscopy [13], energy loss spectroscopy [33], confocal laser scanning microscopy [46], porosimetry [43], and neutron radiography [31]. To address the complexity of adhesive flow through the porous network of interconnecting cell lumens and pits, X-ray tomography has been used for a three-dimensional visualisation of adhesive penetration [16,22,30,32,38]. These studies have provided essential information helping understand the fundamentals of adhesive penetration. However, it has been acknowledged that difficulties arise in establishing direct relationships between adhesive penetration and bond performance due to the many associated parameters, such as the variability of wood species in anatomical characteristics and permeability, the wide variety of adhesive application methods and curing processes, and the many types of adhesive chemistries [14,19,47]. Thus, it is not only difficult to quantitatively

---

* Corresponding author. Tel.: +49 551 3933664; fax: +49 551 399646.
   E-mail address: abastan@gwdg.de (A. Bastani).

http://dx.doi.org/10.1016/j.ijadhadh.2016.04.006

measure penetration but also to isolate its influence on bond quality from the previously mentioned many factors [21].

Modified wood has become a very common and demanded material for different purposes, especially for building, decking and outdoor furniture applications due to its excellent properties, such as enhanced durability, hardness, weathering resistance and dimensional stability [17,25]. On the other hand, the altered structural, chemical and physical properties of wood might influence adhesive penetration and gluing performance of modified wood. The wetting of surfaces of modified wood is known to be affected negatively, and also the various wood modifications have a clear impact on the surface energy characteristics leading to a non-polar character of surfaces [3]. The very low surface polarity of modified wood, the poorer adhesive wetting of the less porous modified wood surfaces, and the fewer chemical bonds between the two surfaces [18] are expected to prevent the formation of attractive polar forces, impair penetration, and accordingly affect negatively the adhesion. However, effective bonding of wood modified with different methods was possible depending on the processing conditions and types of adhesives used [11,26,34,37,41,42]. The reduced shrinking and swelling stresses of modified wood on the cured adhesive bond might also have a positive effect on the bonding performance [37].

Although it is quite essential to investigate the gluing ability of modified wood for developing the market in this area, there are only a few studies that have been carried out on the penetration of adhesives into modified wood. Modification with 10% and 25% phenol-formaldehyde (PF) resulted in a decrease of both phenol resorcinol formaldehyde (PRF) and polyvinyl acetate (PVAc) penetration into beech wood, especially at the highest PF concentration [1]. Nevertheless, the PF modified wood could be bonded satisfactorily with the two adhesive systems under dry conditions with the exception of wood modified with 25% PF solution and PVAc adhesive. For acetylated wood, it has been reported a good degree of penetration for four different gluing systems despite its inferior bonding. Thus, it was confirmed also for modified wood that lumen penetration does not always correlate with bond strength [9]. Heat treatment had no influence on adhesive penetration and there was no relation between bonding performance and penetration [35].

There are two aspects which should be considered while describing adhesive penetration; the first one is the amount of penetrated adhesive into the porous wood structure and the second one is the maximum distance that adhesive particles are able to move inside wood. An anatomical description of penetration mode into different wood tissues is also helpful to better understand the situation under which penetration happens. This study aimed to characterize the penetration behaviour of three conventional gluing systems into furfurylated, N-methylol melamine modified and heat-treated wood after applying pressure by investigating the wood adhesive interfaces with fluorescence microscopy and X-ray micro-computed tomography. The very same adhesives and modified materials were used previously to detect the gross adhesive penetration with no pressure applied [4]. Thus, the relationship between adhesive penetration with and without pressure was also considered. Information on the interaction of gluing systems with these new types of wooden materials can be used to explain the bonding quality of modified wood and, thus, to assure and expand its commercial utilities.

2. Materials and methods

2.1. Wood material and modifications

The wood materials used were boards of furfurylated Scots pine (*Pinus sylvestris* L.), melamine-treated Scots pine, and heat-treated Scots pine and beech (*Fagus sylvatica* L.). Furfurylation was done according to the industrial technology of Kebony ASA (Skien, Norway) using a vacuum and pressure stage followed by curing and vacuum kiln-drying in two concentrations of furfuryl alcohol FA 40 and FA 70, with a 65% and 75% weight percent gain (WPG) respectively. Heat treatment was carried out at 195 and 210 °C by Timura (Holzmanufaktur GmbH, Südharz, Germany) using hot plates with vacuum, press and dewatering stages according to Vacu³ method [44]. Melamine modification was performed using the N-methylol melamine (NMM) resin Madurit MW840/75WA (Ineos Melamines GmbH, Frankfurt, Germany) as an aqueous stock solution with a solid content of approx. 75%. Solutions of 10%, 20% and 30% NMM solid content were prepared, and modification was done by applying vacuum (100 mbar/1 h) and pressure (12 bar/ 2 h) phases in a stainless steel vessel followed by 1 week drying at 120 °C for curing of the NMM resin. The modified boards were conditioned at 20 °C and 65% RH, and then planned samples were prepared with dimensions $100 \times 6 \times 330$ mm^3 (W × T × L).For all used samples, the angle between the growth rings and the surface to be bonded was kept between 30° to 90°.

2.2. Gluing

Three common coldset wood adhesives were selected for this study; two waterborne that cure by solvent evaporation, emulsion polymer isocyanate (EPI) and polyvinyl acetate (PVAc), and a moisture curing one-component polyurethane (PU). All adhesives were provided by Jowat AG (Detmold, Germany) and their details can be seen in Table 1. Using hand brush, the adhesives were applied on the tangential surfaces of the samples at a loading level of 200 g/m^2, and after assembling they were put under a hydraulic press at 1 N mm^{-2} for 4 h at room temperature. Before measurement of adhesive penetration the glued assemblies were left aside for 1 week at room temperature (20 °C). Ten assemblies were used for each adhesive and treatment.

2.3. Microscopy and measurement of penetration

To detect and quantify the radial adhesive penetration into the porous wood structure, 20–30 μm-thick sections were cut using a Reichert-Jung sliding microtome exposing a bondline with a transverse surface at various positions of the specimens. The sections were stained with droplets of 0.5% safranin O solution, and then mounted on glass slides for observation under an Eclipse 50i fluorescence microscope with appropriate filter sets, equipped with a Sight DS-5M-L1 digital camera and a NIS-Elements F

Table 1
Technical information on adhesives.

Property	Adhesive		
	EPI	PU	PVAc
Solids, %	60.0	99.5	49.0
Brookfield viscosity (20 °C), MPa	9.400[a]	10.500	5.000[a]
Density, g/cm^3	1.50	1.15	1.04
pH	7.0	–	5.2

[a] After adding cross linking agent.

software for image analysis (all Nikon, Düsseldorf, Germany). Random areas of penetrated adhesive were used to measure effective penetration (EP) and maximum penetration (MP) according to Sernek et al. [36]. Based on this method, the EP is considered as the total area of adhesive found in the interphase region (volume including both wood cells and adhesive) divided by the width of the bondline. The average of five deepest detected adhesive objects in the interphase region represents MP. With the help of NIS-Elements F software it is possible to separate the bright adhesive objects from the darker background and to calculate the required statistical parameters. For each case, ten regions of interest with dimensions $1100 \times 600 \ \mu m^2$ (W $\times$ H) were selected to measure EP and MP using the following formulas while each obtained value represented the average of 50 measurements:

$$EP = \frac{\sum_i^n A_i}{X_o} \tag{1}$$

where EP = effective penetration, μm; A_i = area of adhesive object i, μm^2; n = number of objects; X_o = width of the maximum rectangle determining the measurement area (1100 μm),

$$MP = \frac{\sum_i^5 (y_i + r_i - y_0)}{5} \tag{2}$$

where MP = maximum depth of penetration, μm y_i = centroid of adhesive object i representing the deepest penetration, μm; r_i = mean radius of adhesive object i, μm; y_o = reference y-coordinate of the bondline interface, μm.

Data on the EP without pressure of the very same adhesives and modified wood materials (e.g. from the same boards, adjacent to the samples of this study), which had been obtained in a previous study by following the same microscopic methodology and had been checked for their significance [4], were used for comparisons with the EP under the hydraulic pressure of this study. For clarification, in the previous study the adhesives had been just spread uniformly on the tangential surfaces of wood samples by hand brushing and samples had been left aside for 1 week at room temperature to let adhesives to penetrate as much as possible.

Visual observation of the modality of penetration of adhesives into the wood structure was also reported for each case to provide a better understanding of the reality of penetration.

2.4. X-ray micro-computed tomography

In both studies with and without pressure, fluorescence microscopy revealed a high amount and depth of penetration of PU adhesive into heat-treated wood also in the form of several isolated adhesive particles with no connection to adjacent cells and quite far from the wood surface. This led to the need of a more detailed three-dimensional (3D) visualisation of penetration of this adhesive into the heat-treated wood structure by means of X-ray micro-computed tomography (XμCT). Scanning of Scots pine heat-treated samples at 195 °C was performed by using a sub-micro-focus XμCT system *nanotom*® s (phoenix|x-ray, GE Sensing & Inspection Technologies GmbH Wunstorf, Germany). The samples were prepared by cutting a small wood section out of the assemblies with a scalpel, which were then vertically placed in a cylindrical glass rod (0.5 cm in diameter) using hot wax. Then, the prepared samples were mounted on the sample rotation stage of the XμCT device. The samples were scanned at a tube voltage of 80 kV, a current of 160 μA, and an X-ray exposure time of 1500 ms for each captured projection. Seven frames were averaged for each single saved radiographic projection and a total of 2000 projections were collected over a 360° range. Based on these settings, the data acquisition took 404 min and a spatial resolution size could reach 1 μm. The principle of XμCT is based on the attenuation of X-ray radiation throughout the sample. The attenuation coefficient

for X-ray is defined by density and elemental composition of the sample material. The different X-ray absorptions are saved in a grayscale-based two-dimensional (2D) projection. Thereby, grayscale values of each pixel characterize the average value of attenuation coefficient. In this method, the dark pixels represent regions within the sample with less X-ray absorption (e.g. air) whereas white or light grey pixels represent dense areas (e.g. wood or glue). Using a set of 2D radiographs, a 3D imaging of the sample volume can be reconstructed, and, in order to achieve a more detailed observation, several sub-volumes were created to choose the best one. Finally, the 3D reconstruction was done using the analysis software *Avizo*® fire 7.1 (VSG, Mériganc Cedex, France).

3. Results and discussion

3.1. Effective penetration (EP) and maximum penetration (MP) in radial direction

Besides the characteristics of the adhesive, penetration is mainly influenced by the structure of wood. Hardwoods have a more complex wood anatomy than softwoods. Beech is a diffuse porous species, which has bigger cellular cavities than Scots pine. Adhesive penetration is mainly dominated by flow through longitudinal vessels in beech and lumens of longitudinal axial tracheids in Scots pine, and it proceeds in paths of lowest resistance towards the wood tissue. Other structural elements such as hardwood fibres are less suitable for adhesive penetration since they have narrow lumens and thicker cell walls that diminish adhesive flow. It can be also expected a slightly higher radial penetration as compared to the tangential one due to the presence of rays with large lumens and thin cell walls and open pit structures running along the radial direction, which might facilitate the flow inside the wood structure [45]. The above considerations help understanding the interaction of adhesives and the modified materials in terms of the specific radial penetration data presented in this study.

Furfurylation significantly decreased the EP of PU adhesive into the wood structure, while no differences were seen in EP with the WPG of furfuryl alcohol (Table 2). In general, this treatment had no significant impact on the EP of EPI and PVAc adhesives. In comparison, the EPI adhesive showed lower EP values than PU and PVAc for both furfurylation treatments (Table 2). This behaviour of EPI adhesive was also reflected in the MP results shown in Fig. 1, and a possible explanation for this behaviour could be the rapid and constant increase of the primary viscosity of the adhesive after mixing with the hardener (Jowat AG, Detmold, Germany). PU adhesive showed the highest MP, for samples treated with lower concentration of furfuryl alcohol (Fig. 1). It can be assumed that filling and blocking wood cells with furfuryl alcohol might have hindered adhesive penetration. Polymerisation of the chemical in the wood structure diminishes the access of wood polymers to water and reduces significantly water uptake, especially at a higher WPG of furfuryl alcohol [3,35,40]. Based on apparent contact angle and surface energy data, surfaces of furfurylated wood and especially the tangential ones show an increased hydrophobicity and a non-polar character.

Melamine treatments also had a negative effect on EP of PU adhesive into the wood structure, with no significant difference in EP, between concentrations of NMM (Table 2). Like furfurylation and with exception of EP of PVAc into samples treated with the highest 30% NMM treatment, this treatment also had no significant impact on EP of EPI and PVAc adhesives and lower EP values were observed for EPI adhesive than PU and PVAc for all concentration of NMM treatments (Table 2). PU adhesive

Table 2
Effective penetration (EP) of EPI, PU and PVAc adhesives under hydraulic pressure towards the radial direction of furfurylated Scots pine (FA 40 and 70), NMM-modified Scots pine (10%, 20% and 30% NMM), and heat-treated Scots pine (HT-S) and beech (HT-B). Mean values and standard deviations in parenthesis[a].

Treatment	EP, µm		
	EPI	PU	PVAc
		Scots pine	
Scots pine-control	15.08 (8.31)ab	86.95 (4.40)a	42.81 (26.23)a
FA 40	23.38 (11.58)a	29.99 (7.80)b	39.36 (21.03)a
FA 70	10.56 (2.95)b	19.05 (4.27)b	21.70 (11.65)a
F-value	5.974[**]	19.786[***]	3.035ns
10% NMM	16.08 (5.80)a	51.82 (15.76)b	52.88 (15.03)a
20% NMM	22.98 (9.62)a	39.33 (21.29)b	39.10 (15.31)ab
30% NMM	14.57 (3.23)a	24.77 (12.32)b	30.48 (10.45)b
F-value	2.985ns	10.119[***]	2.736[*]
HT-S 195 °C	13.11 (4.17)a	75.87 (10.69)a	57.02 (13.27)a
HT-S 210 °C	19.46 (7.16)a	88.96 (12.19)a	21.53 (11.97)b
F-value	2.295ns	0.677ns	9.497[***]
		Beech	
Beech-control	40.78 (13.08)a	76.06 (30.10)a	51.70 (26.66)a
HT-B 195 °C	29.80 (19.77)a	96.87 (28.76)a	83.52 (15.71)b
HT-B 210 °C	32.18 (11.13)a	99.09 (31.41)a	28.97 (13.23)c
F-value	1.459ns	1.779ns	19.882[***]

[ns] differences not statistically significant.

[a] Values followed by a different letter within a column are statistically different (ANOVA and Tukey HSD test).

[*] Differences statistically significant at $P=5\%$.

[**] Differences statistically significant at $P=1\%$.

[***] Differences statistically significant at $P=0.1\%$.

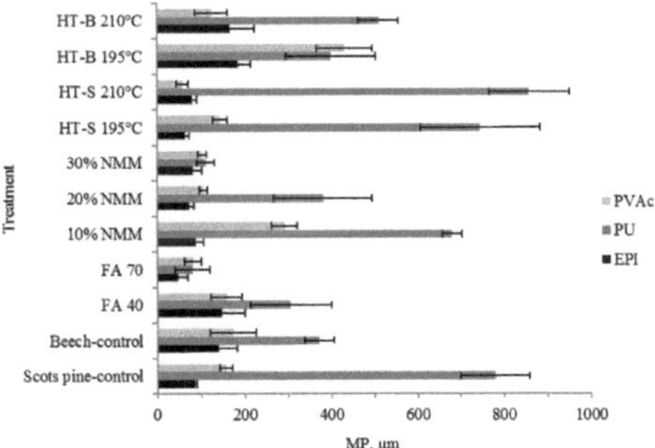

Fig. 1. Maximum penetration (MP) of EPI, PU and PVAc adhesives under hydraulic pressure towards the radial direction of furfurylated Scots pine (FA 40 and 70), NMM-modified Scots pine (10%, 20% and 30% NMM), and heat-treated Scots pine (HT-S) and beech (HT-B). Error bars represent 95% confidence intervals.

represented the highest MP, especially for samples treated with 10 and 20% NMM (Fig. 1). Melamine modification has a similar principle to furfurylation, which is based on filling and blocking wood cells with chemical compounds that may occupy some of the free possible space for penetration of adhesive. In NMM-modified wood, penetration and further polymerisation of the melamine solution into the cell walls brings a bulking effect by creation of covalent bonding between melamine compounds and wood polymers, which decreases the access of water to wood polymers and reduces water uptake [3,27]. (Both tangential and radial NMM-modified wood surfaces revealed an increased hydrophobicity and reduced polarity in a comparable mode to furfurylation with resin load playing no particular role [3] .

Unlike melamine modification and furfurylation, during heat treatment there might be provided more space for further penetration of the adhesive into the wood structure. By increasing the treatment temperature, wood polymers start to decompose and some chemical compounds inside lumens such as extractives may eject from the inner area of wood cells [15,29]. Moreover, heat treatment might cause wood defects, such as cracking and broken cell walls [2,5,6], resulting in the development of additional flow pathways for the adhesive. Heat treatment of Scots pine and beech was found to increase the water uptake in the three main directions of wood, and also affected negatively the wetting and polarity of radial and tangential surfaces [3]. Heat treatment of Scots pine had no influence on the EP of EPI and PU adhesives. In contrast to this, the EP of PVAc decreased significantly into samples treated at 210 °C (Table 2). Similar to furfurylation and NMM modification, EPI adhesive had the lowest EP values as compared to PU and PVAc for both treatment temperatures (Table 2). PU adhesive showed a much deeper penetration into heat-treated Scots pine than two other used adhesives (Fig. 1). Like Scots pine, the heat treatment of beech had no influence on the EP of EPI and PU adhesives (Table 2) and also the EP of PVAc decreased significantly into samples treated at 210 °C. Surprisingly, a significant increase was seen in the EP of this adhesive for beech treated at 195 °C. This finding is in accordance with the decreased contact angle of the water droplets recorded on tangential surfaces of heat-treated beech (195 °C) in a previous study on the same wood materials indicating a more hydrophilic character of the beech after heat treatment [3]. Identical to all modifications, EPI adhesive had the lowest EP values as compared to PU and PVAc for heat-treated beech (Table 2). PU adhesive achieved the highest EP and MP values into heat-treated beech and Scots pine (Table 2, Fig. 1), indicating a large penetrated quantity of PU into heat-treated wood. This fact might be attributed to the low molecular weight of the pre-polymerised PU resin, which allows higher and deeper penetration into the wood structure despite its higher primary viscosity and solid content as compared to the two other adhesives (see Table 1). It should be noted that viscosity is a dynamic state that can drastically change by hardener addition. It has been previously shown that the penetration of an adhesive is inversely related to its molecular weight [10,20]. Another possible explanation for the deep penetration of PU might be the availability of less functional –OH groups in the heat-treated wood, mainly due to large decompositions of hemicelluloses after heat treatment, than in the NMM-treated and furfurylated wood. Therefore, there were less chemical reactions of –OH groups with the isocyanate ingredient of the adhesive, leading to a deeper penetration.

3.2. Comparison of adhesive penetration with and without pressure

As can be seen in Fig. 2, application of pressure provided an altered picture of adhesive penetration in modified wood as compared to the data obtained previously without pressure [4]. Without pressure, the EP was not affected for PU, was reduced in the higher WPG of furfuryl alcohol (FA 70) for EPI and it was increased equally in both FA 40 and 70 treatments for PVAc adhesive. By applying pressure, EP in furfurylated wood was only reduced in the case of the PU adhesive.

The EP of the PU adhesive was lowered in NMM-treated wood both with and without pressure. The EP of EPI was reduced only in the highest NMM treatment (30%) when no pressure was applied, while under pressure no changes were noted. Differences were also noted for the EP of the PVAc adhesive as it was increased in the 10 and 20% NMM-modified wood without pressure but, under pressure, it remained unchanged for these treatments and was even reduced for the 30% NMM modification.

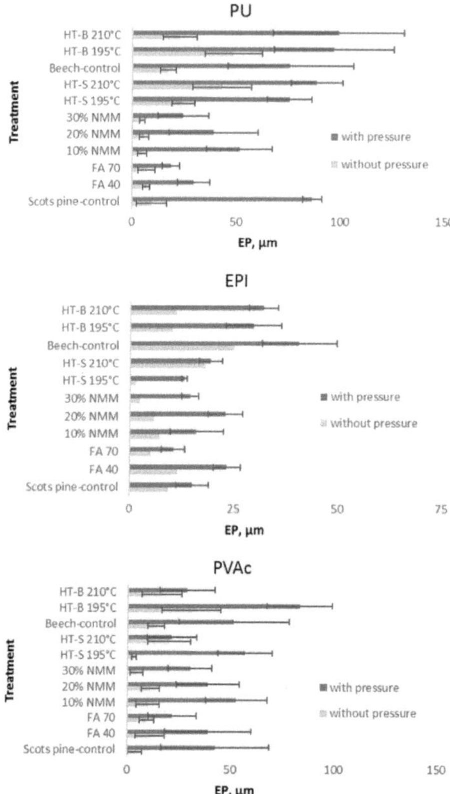

Fig. 2. Penetration of PU, EPI and PVAc adhesives with and without pressure into the radial direction of furfurylated Scots pine (FA 40 and 70), NMM-modified Scots pine (10%, 20% and 30% NMM), and heat-treated Scots pine (HT-S) and beech (HT-B). Error bars represent 95% confidence intervals.

The many changes in EP of all adhesives in heat-treated Scots pine and beech for the measurements without pressure were not seen after pressure. Specifically, without pressure the EP of the PU was increased at both treatment temperatures, the EP of EPI was decreased for the treatment at 195 °C and increased at 210 °C, and the EP of PVAc was increased only at the higher temperature. A variable situation was also seen for data without pressure in beech, where EP was increased only at 195 °C for both PU and PVAc adhesives while it was reduced for EPI after heat treatment. As mentioned earlier, after applying pressure, the heat treatment caused differences only for the EP of the PVAc adhesive, e.g. a decrease for both Scots pine and beech at 210 °C, and an increase for beech at 195 °C.

3.3. Microscopic observations

For PU adhesive, large inter phases were observed in Scots pine controls with distinct and continuous bondlines, containing voids in some parts, and deep penetration of adhesive up to 35 cells at both sides of the bondlines (Fig. 3a). Beech controls showed imperfect and partly starved bondlines that contained several voids as a result of over-penetration of PU adhesive into the wood structure. Interphases were discontinuous and showed a penetration up to 7 fully or partly filled vessels into both sides from the bondlines (Fig. 3b). Reaction of PU adhesive with the moisture content of wood results in the production of CO_2 gas, which causes an inner vapour pressure that drives more adhesive from the bondline towards the wood structure. This phenomenon does not only affect the short molecular chains and lower molecules of PU adhesive bringing a deeper penetration, but also creates starved bondlines and voids, which are typical for this adhesive [16]. In the case of EPI, the Scots pine-control specimens were characterised by a thick and continuous bondline with penetration up to 3 cells in depth. In the beech controls, bondlines were even and rather thin with penetration up to 2 fully filled vessels from the bondline. It is known that PVAc adhesive presents good flow into cell lumens but, given its high molecular weight, most likely it cannot penetrate cell walls [12]. Thick and continuous PVAc bondlines with penetration up to 5 cells in depth were observed for Scots pine controls, while bondlines were thinner for beech controls with penetration up to 3 cells in depth, comprising mostly of full vessels (Fig. 3c).

Samples treated with FA 40 exhibited discontinuous PU bondlines, due to the presence of voids in some parts, with penetration up to 15 cells in depth mainly into the side at the direction of the pressure. Bondlines were thicker in the samples treated with the higher WPG of furfuryl alcohol (FA 70), and penetration was less. In the case of EPI, samples treated with the lower load of furfural alcohol (FA 40) had thick and even bondlines and penetration up to 3 cells in depth, mostly into the application side (direction of applied pressure) of the adhesive (Fig. 3d). Samples treated with the higher WPG of furfulyl alcohol (FA 70) had thinner bondlines with only few penetration areas. For PVAc, thin bondlines with low penetration was observed.

Uneven PU bondlines containing voids in several parts, with a maximum penetration depth of 25 cells in some parts and mainly into the application side of the adhesive was detected for samples treated with the lowest concentration (10%) of NMM. The bondlines for samples treated with 20% NMM were quite distinct with penetration up to 15 cells in depth, while for the 30% NMM-modified samples bondlines were thinner and with less penetration. For EPI adhesive, NMM-modified wood had thin and continuous bondlines with limited, mostly 1-layer cell penetration.

Samples treated with the lowest concentration of NMM (10%) presented imperfect PVAc bondlines with penetration up to 5 partly and fully filled cells in depth, and mostly in the application side of the adhesive. Less penetration was seen into the 20 and 30% NMM treatments, and bondlines were very thin for the samples treated with the highest NMM concentration (Fig. 3e).

In the case of heat-treated Scots pine at 195 °C, distinct and almost continuous PU bondlines with large amounts of penetrated adhesive mainly in the application side of adhesive were found (Fig. 3f). In comparison, bondlines were thicker for Scots pine samples treated at 210 °C. Most of the cells were fully filled with the exception of few earlywood cells partly filled with the adhesive. For both treatment temperatures, the maximum depth of penetration was up to 30 cells presenting large interphase regions. In the case of EPI adhesive, the heat-treated Scots pine had thin and continuous bondlines with limited, mostly 1-layer cell penetration while for PVAc, penetration into heat-treated Scots pine appeared as large areas of penetrated adhesive around thin bondlines for samples treated at 195 °C, and somewhat less penetration characterised the samples treated at 210 °C.

Large quantities of PU adhesive were also penetrated into both sides from the bondlines of heat-treated beech (Fig. 3g). The

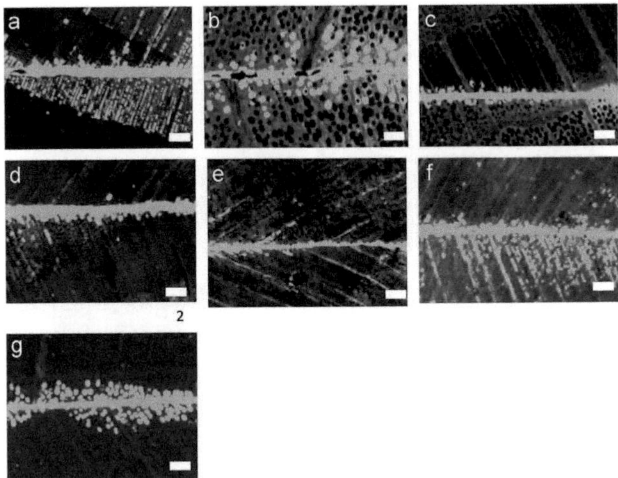

Fig. 3. Fluorescence photomicrographs with transverse view of horizontal bondlines showing the radial penetration of adhesives. PU adhesive in Scots pine (a) and beech control (b), PVAc adhesive in beech control (c), EPI adhesive in furfurylated FA 40 Scots pine (d), PVAc adhesive in 30% NMM Scots pine (e), PU in heat-treated Scots pine (f) and beech (g) at 195 °C. Scale bars=200 μm. Note: in each photomicrograph, the lower part of the wood–adhesive interface from the bondlines represents the side the pressure was applied.

resulting bondlines were intact with penetration up to 7 cells in depth in some parts, including mostly fully filled vessels. A few tiny voids were only seen within the bondlines of beech samples treated at 210 °C. For EPI adhesive, in the case of heat-treated beech at 195 °C, distinct bondlines with maximum penetration up to 3 fully filled vessels in depth were observed in some parts of the interphase. Bondlines were very thin for beech treated at 210 °C, and most of the penetration occurred into the application side of the adhesive. Very thin PVAc bondlines, with very large penetration areas were observed for beech treated at 195 °C. Penetration occurred in both sides from the bondlines, presenting large interphases. Less penetration, mostly 1-cell-layer in depth, was identified for beech treated at 210 °C.

3.4. X-ray micro-computed tomography

Fig. 4a–f illustrates various 3D views of penetration of PU adhesive into the micron scale structure of heat-treated Scots pine wood at 195 °C. In Fig. 4a–d, the dark colour is assigned for air (voids) inside the sub-volume where the high X-ray density parts such as cell walls and adhesive are indicated by light grey or white. Having a similar X-ray density, the grey-scale values of cell walls and adhesive partly overlapped, hence they could not be perfectly segmented. Nevertheless, it was still possible to distinguish between wood and adhesive parts. As described before, the photomicrographs with fluorescence microscopy showed a high amount and depth of penetration of PU adhesive into heat-treated wood, even up to 25 cells from the bondline in some parts. Also, several isolated adhesive parts without any connection to other filled cells around them were detected. Due to limitations of the 2D fluorescence microscopy images, it was not possible to explain this penetration behaviour without examination of the 3D state by using XμCT. Fig. 4b is a 3D rendering of the same 2D image of Fig. 4a with a better visual contrast between wood and adhesive parts. Both images represented the path of penetration of adhesive

from the top of the wood surface (left side of image) through a big ray (in the middle of the image) into surrounding pits and lumens of adjacent axial tracheids. In Fig. 4c–d it was attempted to depict the movement of adhesive in different selected areas (1–7) of a chosen sub-volume. The pattern of flow of adhesive in the radial direction, along a ray and into adjacent axial tracheids, is clearly shown in Fig. 4e. A similar radial flow of phenol-formaldehyde adhesive has been reported for Douglas-fir [30]. Fig. 4f illustrates the flow pattern of adhesive by removing the wood substance and in presence of air.

4. Conclusions

The EP and MP of EPI, PVAc and PU adhesives under hydraulic pressure into different types of modified wood were examined by fluorescence microscopy and the results were compared to a pervious study without pressure [4]. The results of penetration with pressure showed that furfurlylation and NMM modification had a negative influence on EP of PU into modified wood. EP of PVAc only decreased after 30% NMM treatment. After heat treatment, the EP of PVAc into Scots pine and beech wood treated at 210 °C decreased considerably. In contrast, the EP of this adhesive increased into heat-treated beech wood at 195 °C. In the case of furfurylation, a deeper penetration of all adhesives was recorded into wood treated with lower concentration of furfuryl alcohol. PU represented a much deeper penetration into NMM-modified and heat-treated wood than the two other glue types with the exception of beech wood treated at 195 °C. Besides the quantitative measurements, the described analysis of fluorescence photomicrographs provided more detailed information on the modality of adhesive penetration into the modified wood tissue. Application of pressure gave a rather different behaviour of adhesive penetration in modified wood as compared to the previous data on penetration without pressure. The 3D visualisation of penetration

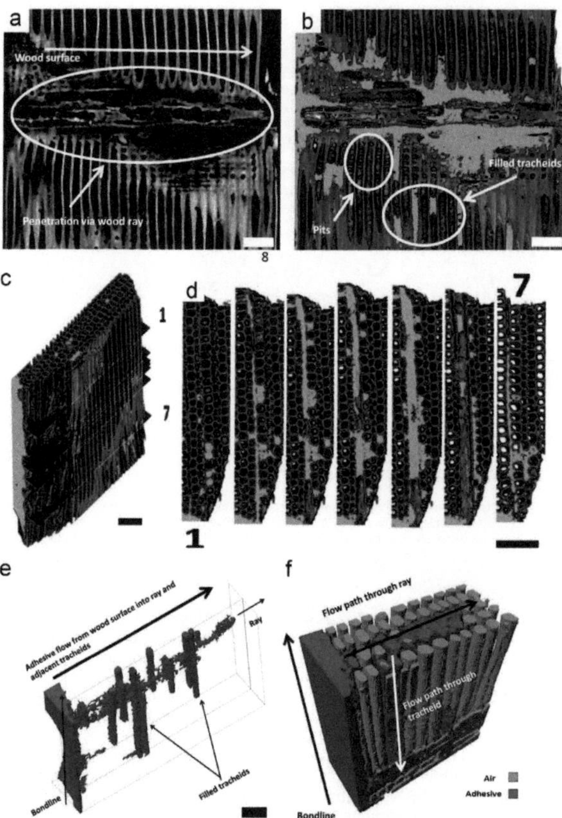

Fig. 4. 3D visualisation of penetration of PU adhesive into heat-treated Scots pine at 195 °C as captured by XμCT. Penetration route from the surface into the wood tissue (left to right) (a, b) and into different areas (1–7) of a selected sum-volume (c, d); flow pattern of penetration in radial direction (e, f). Scale bars = 100 μm.

of PU adhesive into heat-treated Scots pine was also investigated. The 3D pattern obtained by XμCT of PU adhesive flow in the radial direction of heat-treated Scots pine provided an understanding of the pathways this adhesive used (e.g. pits and lumens of adjacent axial tracheids) to penetrate wood.

References

[1] Adamopoulos S, Bastani A, Gascon-Garrido P, Militz H, Mai C. Adhesive bonding of beech wood modified with a phenol formaldehyde compound. Eur J Wood Prod 2012;70(6):897–901.

[2] Altgen M, Adamopoulos S, Militz H. Wood anatomical changes during industrial-scale production of thermally-modified Norway spruce and Scots pine. Wood Mater Sci Eng 2015. http://dx.doi.org/10.1080/17480272.2014.988750.

[3] Bastani A, Adamopoulos S, Militz H. Water uptake and wetting behaviour of furfurylated, N-methylol melamine modified and heat-treated wood. Eur J Wood Prod 2015;73(5):627–34.

[4] Bastani A, Adamopoulos S, Militz H. Gross adhesive penetration in furfury-lated, N-methylol melamine-modified and heat-treated wood examined by fluorescence microscopy. Eur J Wood Prod 2015;73(5):635–42.

[5] Boonstra MJ, Rijsdijk JF, Sander C, Kegel E, Tjeerdsma B, Militz H, et al. Microstructural and physical aspects of heat treated wood. Part 1 softwoods. Maderas Cienc Tecnol 2006;8:193–208.

[6] Boonstra MJ, Rijsdijk JF, Sander C, Kegel E, Tjeerdsma B, Militz H, et al. Microstructural and physical aspects of heat treated wood. Part 2 hardwoods. Maderas Cienc Tecnol 2006;8:209–17.

[7] Brady DE, Kamke FA. Effect of hot-pressing parameters on resin penetration. For Prod J 1988;38(11/12):63–8.

[8] Bolton AJ, Dinwoodie JM, Davies DA. The validity of the use of SEM/EDAX as a tool for the detection of UF resin penetration into wood cell walls in particleboards. Wood Sci Technol 1988;22(4):345–56.

[9] Chandler JG, Brandon LR, Frihart CR. Examination of adhesive penetration in modified wood using fluorescence microscopy. In: ASC spring convention and exposition. Columbus, OH; April 17–20, 2005. p. 10.

[10] Frazier CE, Schmidt R, Ni J. Towards a molecular understanding of wood-isocyanate adhesive bondline. In: Proceedings of the 3rd Pacific rim bio-based composite symposium. Kyoto, Japan; Dec 2–5, 1996. p 383–91.

[11] Frihart CR, Brandon R, Ibach RE.Selectivity of bonding for modified wood. In: Proceedings, 27th annual meeting of The adhesion Society, Inc. Wilmington, NC, February 15-18, 2004, pp. 329-31.

[12] Frihart CR. Wood adhesion and adhesives. In: Handbook of wood chemistry and wood composites. Boca Raton, Fla: CRC Press; 2005. p. 215–78.

[13] Gindl W, Dessipri E, Wimmer R. Using UV-microscopy to study diffusion of melamine-urea-formaldehyde in cell walls of spruce wood. Holzforschung 2002;56(1):103–7.

[14] Gavrilovic-Grmusa I, Dunky M, Miljkovic J, Djiporovic-Momcilovic M. Influence of the viscosity of UF resins on the radial and tangential penetration into poplar wood and on the shear strength of adhesive joints. Holzforschung 2012;66(7):849–56.

[15] Hakkou M, Pétrissans M, El Bakali I, Gérardin P, Zoulalian A. Wettability changes and mass loss during heat treatment of wood. Holzforschung 2005;59:35–7.

[16] Hass P, Wittel FK, Mendoza M, Herrmann HJ, Niemz P. Adhesive penetration in beech wood: experiments. Wood Sci Technol 2012;46(1):243–56.

[17] Hill CAS. Wood modification: chemical, thermal and other processes. Chichester, UK: John Wiley & Sons; 2006.

[18] Hunt CG, Brandon R, Ibach RE, Frihart CR. What does bonding to modified wood tell us about adhesion? In: Proceedings 5th COST E34 international workshop: bonding of modified wood. Slovenia: Bled; 2007.pp .47-56.

[19] Johnson SE, Kamke FA. Quantitative analysis of gross adhesive penetration in wood using fluorescence microscopy. J Adhes 1992;40:47–61.

[20] Johnson SE, Kamke FA. Characteristics of phenolformaldehyde adhesive bonds in steam injection pressed flakeboard. Wood Fiber Sci 1994;26(2):259–69.

[21] Kamke FA, Lee JN. Adhesive penetration of wood-a review. Wood Fiber Sci 2007;39(2):205–20.

[22] Kamke FA, Nairn JA, Muszynski L, Paris JL, Schwarzkopf M, Xiao X. Methodology for micromechanical analysis of wood adhesive bonds using X-ray computed tomography and numerical modelling. Wood Fiber Sci 2014;46 (1):15–28.

[23] Konnerth J, Harper D, Lee SH, Rails TG, Gindl W. Adhesive penetration of wood cell walls investigated by scanning thermal microscopy (SThM). Holzforschung 2008;62:91–8.

[24] Koran Z, Vasishth RC. Scanning electron microscopy of plywood glue lines. Wood Fiber 1972;3(4):202–9.

[25] Kumar S. Chemical modification of wood. Wood Fiber Sci 1994;26(2):270–80.

[26] Kurt R, Mai C, Krause A, Militz H. Hydroxymethylated resorcinol (HMR) priming agent for improved bondability of silicone modified wood glued with a polyvinyl acetate adhesive. Holz Roh Werkst 2008;66:305–7.

[27] Lukowsky D. Holzschutz mit Melaminharzen. Germany: PhD Thesis, University of Hamburg; 1999.

[28] Marra A. Technology of wood bonding principles in practice. New York: Van Nostrand Reinhold; 1992.

[29] Militz H. Thermal treatment of wood: european processes and their background. International Research Group on Wood Preservation. IRG/WP 02-40241; 2002.

[30] Modzel G, Kamke FA, De Carlo F. Comparative analysis of a wood: adhesive bondline. Wood Sci Technol 2011;45(1):147–58.

[31] Niemz P, Mannes D, Lehmann E, Vontobel P, Haase S. Untersuchungen zur Verteilung des Klebstoffes im Bereich der Leimfuge mittels Neutronenradiographie und Mikroskopie. Holz Roh Werkst 2004;62(6):424–32.

[32] Paris JL, Kamke FA, Mbachu R, Kraushaar Gibson S. Phenol formaldehyde adhesives formulated for advanced X-ray imaging in wood-composite bondlines. J Mater Sci 2013;49(2):580–91.

[33] Rapp AO, Bestgen H, Adam W, Peck RD. Electron energy loss spectroscopy (EELS) for quantification of cell-wall penetration of a melamine resin. Holzforschung 1999;53(2):111–7.

[34] Sahin Kol H, özbay G, Altun S. Shear strength of heat-treated tali (Erythrophleum ivorense) and iroko (Chlorophora excelsa) woods, bonded with various adhesives. BioResources 2009;4(4):1545–54.

[35] Sernek M. Comparative analysis of inactivated wood surfaces (PhD dissertation). Blacksburg, VA: Virginia Polytechnic Institute and State University; 2002. p. 179.

[36] Sernek M, Resnik J, Kamke FA. Penetration of liquid urea-formaldehyde adhesive into beech wood. Wood Fiber Sci 1999;31(1):41–8.

[37] Sernek M, Boonstra M, Pizzi A, Despres A, Gérardin P. Bonding performance of heat treated wood with structural adhesives. Holz Roh Werkst 2008;66:173–80.

[38] Shaler SM, Keane DT, Wang H, Mott L, Landis E, Holtxman L. Microtomography of cellulosic structures. In: Proceedings of the TAPPI, Process Product, date, city; 1998. p. 89-46.

[39] Smith LA, Côté WA. Studies of penetration of phenol- formaldehyde resin into wood cell walls with the SEM and energy-dispersive X-ray analyser. Wood Fiber 1971;3(1):56–7.

[40] Treu A, Pilgård, Puttmann S, Krause A, Westin M (2009) Material properties of furfurylated wood for window production. Proceedings of The International Research Group on Wood Protection, 40th Annual Meeting, 24-28 May, Beijing, China. IRG/WP 09-40480.

[41] van der Zee ME, Schipholt NL, Tjeerdsma BF, Brynildesn P, Mohoric I. Glueability and paintability of furfurylated wood (Kebony). In: Hill CAS, editor. Proceedings of the third european conference on wood modification. Cardiff, UK; 2007. p. 231–4.

[42] Vick CB, Rowell RM. Adhesive bonding of acetylated wood. Int J Adhes Adhes 1990;10(4):263–72.

[43] Wang WQ, Yan N. Characterising liquid resin penetration in wood using a mercury intrusion porosimeter. Wood Fiber Sci 2005;37(3):505–14.

[44] Wetzig M, Niemz P, Sieverts T, Bergemann H. Mechanische und physikalische Eigenschaften von mit dem Vakuumpress-Trocknungsverfahren thermisch behandeltem Holz. Bauphysik 2012;34(1):1–10.

[45] Weigenand O, Militz H, Tingaut P, Sèbe G, De Jeso B, Mai C. Penetration of amino-silicone micro- and macro-emulsions into Scots pine sapwood and the effect on water-related properties. Holzforschung 2007;61(1):51–9.

[46] Xing C, Riedl B, Cloutier A, Shaler SM. Characterisation of urea-formaldehyde resin penetration into medium density fibreboard fibers. Wood Sci Technol 2005;39(5):374–84.

[47] Zheng J, Fox SC, Frazier CE. Rheological wood penetration, and fracture performance studies of PF/pMDI hybrid resins. Prod J 2004;54(10):74–81.

Paper 5

WOOD MATERIAL SCIENCE & ENGINEERING, 2016
http://dx.doi.org/10.1080/17480272.2016.1164754

Shear strength of furfurylated, N-methylol melamine and thermally modified wood bonded with three conventional adhesives

Alireza Bastani[a], Stergios Adamopoulos[b] and Holger Militz[a]

[a]Wood Biology and Wood Products, Burckhardt Institute, Georg-August-University Göttingen, Göttingen, Germany; [b]Department of Forestry and Wood Technology, Linnaeus University, Växjo, Sweden

ABSTRACT

The shear strength of furfurylated, N-methylol melamine (NMM) and thermally modified wood bonded with emulsion polymer isocyanate, polyvinyl acetate (PVAc), and polyurethane (PU) adhesives was examined. Furfurylation and NMM modification of Scots pine had a significant negative effect on the bonding strength with all adhesives irrespective of the treatment intensity. The obtained low-shear strength values were related to the brittle nature of the wood after modifications rather to the failure of the bondline. PVAc showed a better bonding performance with both furfurylated and NMM modified wood while the combination of furfurylated wood and PU gave the highest reduction in bonding strength (47–51%). Shear strength also decreased significantly after thermal modification in both Scots pine (36–56%) and beech (34–48%) with all adhesives. With the exception of thermally modified beech samples bonded with PU, bondline was found to be the weakest link in thermally modified wood as it was revealed by the wood failure surfaces. Bondline thickness and effective penetration of adhesives did not relate to the shear strength of all modified wood materials. The lower shear strength of modified wood could be attributed to other factors, such as the reduced chemical bonding or mechanical interlocking of adhesives, and the reduced strength of brittle modified wood substrate.

ARTICLE HISTORY
Received 30 May 2015
Revised 8 March 2016
Accepted 8 March 2016

KEYWORDS
Shear strength; modified wood; adhesive; bondline thickness; effective penetration

Introduction

Wood construction and furniture industries are highly dependent on the quality of glued-wood structures and elements. The durability of bondlines has a determinant influence on the ultimate performance of the wood-adhesive joints, and hence of the whole wooden structure (Modzel et al. 2011). As suggested by Marra (1992), an adhesive bond can be schemed as a chain-link including nine attached links where the maximum strength of the bondline is determined by the weakest link of the chain. In this scheme, link 1 represents the pure adhesive part which is located in middle of the bondline, links 2–3 exhibit the nonhomogeneous and partly cured boundary layer of adhesive, and links 4–5 stand between this boundary layer and the wood material. These later two links form the adhesion mechanism, for example, mechanical interlocking, covalent bonding, or secondary chemical bonds. Links 6–7 represent the wood substrate affected by the wood surface preparation or bonding processes, and links 8–9 represent the non-interacted wooden material. Penetration of adhesive affects links 4–7, and consequently all possible adhesion mechanisms (Kamke and Lee 2007). A proportional amount of penetrated adhesive into the wood structure is needed for both adequate bond performance and optimum adhesive costs. A low quantity of penetrated adhesive (under-penetration) or excessive penetration (over-penetration) provides limited surface contact with the interior surfaces for chemical bonding or mechanical interlocking (Johnson and Kamke 1994).

Wood modification is an environmental friendly technology aiming to improve the biological durability, dimensional stability, and hardness of wood (Hill 2006). On the other hand, after modification wood undergoes physical and chemical changes, which might potentially affect its bonding properties. For example, the creation of a less polar and less porous wood surface after modification results in the formation of less free -OH groups for bonding, and hence to the establishment of weaker bonds due to the poorer wetting of the wood surface (Hunt et al. 2007). Adamopoulos et al. (2012) reported a significant reduction in shear strength and wood failure percentage after modification of beech with phenol formaldehyde (PF) for polyvinyl acetate (PVAc) bonded samples treated with 25% PF solution. It was also found a lower penetration of the used adhesives after PF modification especially at the higher load treatment solution of 25%. Dry tensile strengths of montan ester wax-treated pine wood glued with PVAc and melamine formaldehyde adhesives were in a comparable range to those of untreated wood, while under wet conditions the strength values were significantly lower (Kurt et al. 2008). The effect of acetylation on the bonding performance of wood glued with various adhesives was investigated in several studies with diverse results depending on the adhesive systems used (Vick and Rowell 1990, Frihart et al. 2004, Brandon et al. 2005). It was stated that good lumen penetration did not cause a desirable bonding strength for acetylated wood (Chandler et al. 2005). Acceptable bonds were provided for furfurylated wood when emulsion polymer

CONTACT Alireza Bastani ✉ abastan@gwdg.de

isocyanate (EPI) and melamine urea formaldehyde were used for adhesion, despite the detected differences in the glue line width among different wood species (van der Zee *et al.* 2007). Heat treatment was reported to affect negatively the bonding strength, mainly for two reasons; the low pH value and the low wettability of heat-treated wood as compared to untreated (Sernek *et al.* 2008). Heat treatment was also found to decrease the shear strength of tali and iroko wood bonded with different adhesives (Sahin Kol *et al.* 2009).

As modified wood has become a more used material for building and furniture applications during the last years, it is essential to expand the knowledge on its bonding behaviour. This can be achieved by providing suggestions for effective bonding of different modified wood materials and by understanding better the associated mechanisms and interaction with different adhesives. The present study aimed to investigate the shear strength of furfurylated, N-methylol melamine (NMM) and thermally modified wood bonded with three common wood adhesives. It was also inquired the relationship between shear strength and (a) bondline thickness and (b) adhesive penetration. Data on adhesive penetration were available from a previous study on the very same wood materials and adhesives (Bastani *et al.* 2015a).

Materials and methods

Wood materials and modifications

The modified wood materials used were furfurylated and melamine modified Scots pine (*Pinus sylvestris* L.), and thermally modified Scots pine and beech (*Fagus sylvatica* L.) with primary board dimensions of $1400 \times 100 \times 30$ mm^3 ($L \times W \times T$). Furfurylation was carried out by Kebony ASA (Skien, Norway) through an industrial process with furfuryl alcohol (FA) 40 and 70 concentrations of FA, which are commercial solutions of Kebony ASA, with a 65% and 75% weight per cent gain (WPG) respectively. Using NMM resin Madurit MW840/75WA (Ineos Melamines GmbH, Frankfurt, Germany), melamine modification was done in a stainless steel vessel by applying 10%, 20% and 30% NMM solid content solutions through a full cell process, including an initial vacuum phase of 100 mbars for 1 h, a pressure phase of 12 bars for 2 h, and curing of the resin at 120°C for 1 week in a drying kiln at the end. Thermal modification was performed by Timura Holzmanufaktur GmbH (Südharz, Germany) at 195°C and 210°C through an industrial scale vacuum-press dewatering method (Vacu^3). All the modified boards were conditioned in a climate chamber at standard conditions (20°C/65% RH) to reach equilibrium moisture content, and then planned samples were prepared with dimensions $330 \times 100 \times 6$ ($L \times W \times T$) mm^3 for further testing. The angle between the growth rings and the surface to be bonded was kept between 30° and 90°. These samples were part of collectives providing material for measuring adhesive penetration in a previous study (Bastani *et al.* 2015a). In each collective, four specimens were always taken in one longitudinal sequence to minimize the influence of natural property variation.

Gluing and tensile strength testing

EPI andPVAc as waterborne adhesives curing by solvent evaporation, and a moisture curing one-component polyurethane (PU) adhesive (all provided by Jowat AG, Detmold, Germany) were applied on the wood samples by using a hand brush. Information on the adhesives is provided in Table 1. The prepared assemblies were left aside for 1 week at room temperature (20°C) in order to let adhesives to cure and penetrate sufficiently, and then they were put under a hydraulic press at 1 N mm^{-2} for 4 h at room temperature. Before testing, all assemblies were conditioned at 20°C and 65% RH in a climate chamber. Measurement of tensile shear strength was carried out in a Zwick/Roell Z010 universal testing machine with specimen geometries according to the European standard EN 302-1 (1992). For each adhesive and treatment, 10 replicates were tested.

Observation of wood-adhesive bondline

After the shear strength tests, 20–30 μm thick sections exposing a bondline with a cross-sectional surface at various longitudinal positions of the assemblies were prepared using a Reichert-Jung sliding microtome to measure the bondline thickness (adhesive part between the two adherents). The sections were stained with a safranin O solution and were mounted on glass slides. Then, the prepared slides were observed under an Eclipse 50i fluorescence microscope equipped with appropriate filter sets, a Sight DS-5M-L1 digital camera, and NIS-Elements F software for image analysis (all Nikon, Düsseldorf, Germany). For each case, 10 randomly measurements were performed across each bondline with a fixed measurement interval.

Simple Pearson correlation was used to determine the relation between shear strength and bondline thickness. Shear strength was also correlated to the effective penetration (EP) of the adhesives. These data had been determined previously by Bastani *et al.* (2015a) following the definitions found in Sernek *et al.* (1999) on assemblies of the very same modified materials (e.g. from the same boards, adjacent to the samples of this study) and adhesives by applying the same gluing and pressure method and parameters.

Results and discussion

Shear strength

The shear strength results and the corresponding wood failures are shown in Table 2. Furfurylation and NMM modification significantly reduced the shear strength of Scots pine for all three adhesives used (ANOVA and Tukey HSD test, $P = 5\%$). With the exception of FA 70 treated wood bonded with PVAc, which showed the lowest but still significant reduction in shear strength (16%), in all other cases the shear strength values were drastically dropped by 28–51% as compared to the untreated controls. The treatment intensity, for example, the load of FA and the concentration of the NMM solution, had no significant

Table 1. Characteristics of the adhesives used in the study.

Property	Adhesive		
	EPI	PU	PVAc
Solids, %	60	99.5	49
Brookfield viscosity (20°C), MPa	9400[a]	10,500	5000[a]
Density, g/cm^3	1.5	1.15	1.04
pH	7	–	5.2

[a]After adding cross-linking agent.

effect in the shear strength reduction. Again, the exception was the furfurylated FA 70 wood bonded with PVAc, which provided significant better bonds than the treatment with the lower WPG of FA 40. Both furfurylated and NMM modified wood could be bonded better with the PVAc adhesive as they showed the lowest reduction in shear strength, for example, 16–31% for furfurylated wood and 28–33% for NMM modified wood. In comparison, EPI and PU adhesives had a similar high loss of shear strength for furfurylated (36–45% for EPI) and NMM modified wood (35–51% for PU). The worst bonding performance was noted for furfurylated wood and PU adhesive as it lost 47–51% of its shear strength.

It could be assumed that penetration and polymerization of FA into the wood cell walls (Stamm 1977) and incorporation of the 3D cross-linked melamine resin in the wood matrix (Mahnert et al. 2013, Kielmann et al. 2014) might have hindered chemical bonding or mechanical interlocking of adhesives in the modified wood. Surface repellancy could also be a contributory factor to shear strength decrease. Sessile drop apparent contact angles for water and diiodomethane and corresponding surface energies on tangential and radial wood surfaces revealed an increased hydrophobicity and reduced polarity of NMM modified and furfurylated wood (Bastani et al. 2015b). However, by looking the wood failure data (Table 1) and surfaces (Figure 1), a different explanation can be given for the lower shear strength of furfurylated and NMM modified wood. A very high wood failure percentage of 90–100% was seen for NMM modified wood accompanied with a strong grooved surface due to its brittle nature (Figure 1(a)). A similar behaviour was observed in most of the cases for furfurylated wood with high wood failure percentages (80–95%) at comparable levels to controls (85–95%) and grooved surface appearance (Figure 1(b)). An exception was noted for the rather lower failure of 70% for FA 40 modified wood glued with PU adhesive. It was thus implied that the bondline was stronger compared to the furfurylated and NMM modified wood itself, and therefore the reduction in shear strength was mainly due to the reduction of wood strength. For furfurylation, it was found that the bondline was the weakest link for beech and maple, where in southern yellow pine the bondline was stronger compared to the wood itself (van der Zee et al. 2007). Despite a slight reduction in dry shear strength, furfurylated wood could be glued satisfactorily with EPI or melamine urea formaldehyde adhesives (van der Zee et al. 2007). That was also true for other modifications leading to cell wall bulking depending on the adhesive system used (Vick and Rowell 1990, Adamopoulos et al. 2012). However, as also indicated in this study, the brittleness and reduced strength of furfurylated and

melamine modified wood (Stig Lande et al. 2004, Kielmann et al. 2013) should be considered for its gluing.

Thermal modification also resulted to a significant reduction in shear strength for all adhesives at comparable to furfurylation and NMM modification high levels of 35–56% for Scots pine and 34–48% for beech (ANOVA and Tukey HSD test, $P = 5\%$). Especially for beech, the shear strength did not reach the minimum required value of 10 Nmm^{-2} allocated by the European Standard EN 301 (2006) for the dry tensile strength of unmodified wood. PU adhesive performed slightly better in gluing thermally modified beech than the other adhesives, and the obtained values were very close to the required level of shear strength mentioned in the standard. Sahin Kol et al. (2009) also reported a good performance of PU adhesive for gluing thermally modified wood. The treatment temperature affected significantly the bonding strength of Scots pine and beech only in the case of EPI adhesive. With this adhesive, Scots pine wood could be bonded better at the higher temperature of 210°C while the opposite occurred for beech. For 195°C treatment temperature, all adhesives performed equally for Scots pine reducing the shear strength by 48–56%, and for beech EPI provided a better bonding showing the lowest reduction (34%) in shear strength than the two other adhesives. Based on the shear strength loss, Scots pine and beech wood treated at 210°C could be bonded better by PVAc and PU, respectively.

The negative effect of thermal modification on bonding strength was also reported by Sahin Kol et al. (2009) by testing different adhesives. On the contrary, Sernek et al. (2007) showed that thermally modified wood could be bonded satisfactorily with hot curing adhesives with the treatment temperature having no effect on the shear strength. As shown from the reduced wood failure percentages of both Scots pine and beech after thermal modification (Table 2),

Table 2. Shear strength and wood failure (WF) percentage of furfurylated (FA 40 and 70) and NMM modified Scots pine (10%, 20% and 30% NMM), and thermally modified Scots pine (HT-S) and beech (HT-B) assemblies bonded with EPI, PU, and PVAc adhesives.

Treatment	Shear strength, Nmm^{-2}			WF, %		
	EPI	PU	PVAc	EPI	PU	PVAc
Scots pine						
Scots pine-control	10.44 (1.7)a	11.00 (2.0)a	9.12 (1.0)a	85	90	95
FA 40	6.69 (0.2)b	5.39 (0.5)b	6.27 (0.5)b	95	70	90
FA 70	5.73 (0.8)b	5.87 (0.3)b	7.69 (0.9)c	80	90	90
F-value	28.363*	33.253*	14.480*			
10% NMM	6.22 (0.1)b	7.15 (1.1)b	6.14 (0.7)b	100	95	90
20% NMM	5.99 (1.2)b	6.84 (1.3)b	6.28 (0.5)b	100	100	95
30% NMM	5.95 (0.5)b	6.48 (0.7)b	6.61 (0.6)b	90	100	95
F-value	20.526*	13.474*	19.769*			
HT-S 195°C	4.56 (0.5)b	5.70 (0.3)b	4.40 (0.9)b	90	75	85
HT-S 210°C	6.41 (0.9)c	4.88 (0.7)b	5.89 (1.2)b	80	80	85
F-value	35.512*	36.544*	9.146*			
Beech						
Beech-control	14.49 (1.5)a	15.94 (0.8)a	15.32 (1.1)a	90	100	95
HT-B 195°C	9.60 (1.6)b	9.70 (1.6)b	8.14 (1.0)b	80	100	85
HT-B 210°C	7.47 (0.6)c	9.86 (1.5)b	8.02 (1.6)b	85	100	85
F-value	39.991*	34.435*	48.768*			

Notes: Mean values and standard deviations in parentheses. Values followed by a different letter within a column are statistically different (ANOVA and Tukey HSD test).
*Differences statistically significant at $P = 5\%$

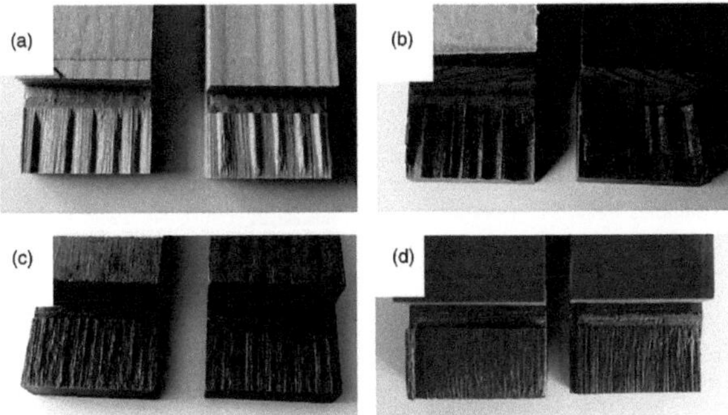

Figure 1. Wood failure surfaces of 10% NMM modified Scots pine bonded with EPI (a), FA 70 modified Scots pine bonded with PVAc (b), thermally modified beech at 210°C bonded with PU (c), and thermally modified Scots pine at 195°C bonded with EPI (d).

the bondline was the weakest link. An exception was seen for beech bonded with PU where for the untreated and thermally modified samples the bondline was found stronger compared to the wood itself with wood failure percentage of 100% (see also Figure 1(c)). The grooved surface of the furfurylated and mostly the NMM modified test samples was not observed for the thermally modified ones (Figure 1(c) and 1(d)). One of the main reasons for the decreased adhesion is the existence of the less polar groups for bonding in thermally modified wood (Inari *et al.* 2007, Sernek *et al.* 2008). The poor bonding of thermally modified wood, as well the furfurylated and NMM modified, could be also explained by its poorer wettability (Bastani *et al.* 2015b), which could also hinder the proper curing of adhesives (Boonstra *et al.* 1998).The reduced wettability could potentially have an adverse effect on the bonding strength of furfurylated and NMM modified wood, but as explained earlier it might have been offset by the lower strength of the brittle modified wood and by the reduced internal surfaces for chemical bonding or mechanical interlocking of adhesives in the bulked modified wood.

Shear strength vs. adhesive penetration and bondline thickness

Penetration of adhesives into the porous wood structure is proposed as one of the factors influencing bonding performance (Brady and Kamke 1988, Marra 1992, Kamke and Lee 2007). For example, an increased penetration of adhesive in southern pine plywood caused improvements in shear strength and wood failure percentage (Hse 1971), and also a higher bond strength was attributed to an increase of the degree of penetration by decreasing adhesive's viscosity (Gavrilovic-Grmusa *et al.* 2012). On the other hand, several studies indicated that a good adhesive penetration does not always relate to improved shear strength and wood failure percentage values due to the complicated structure of wood and other factors involved in the ultimate bonding performance (Suchsland 1958, Chandler *et al.* 2005). In our previous study on the same materials (Bastani *et al.* 2015a), furfurlylation and NMM modification had a negative effect on EP of PU adhesive where for PVAc, this negative effect was seen in the case of 30% NMM treatment and thermal

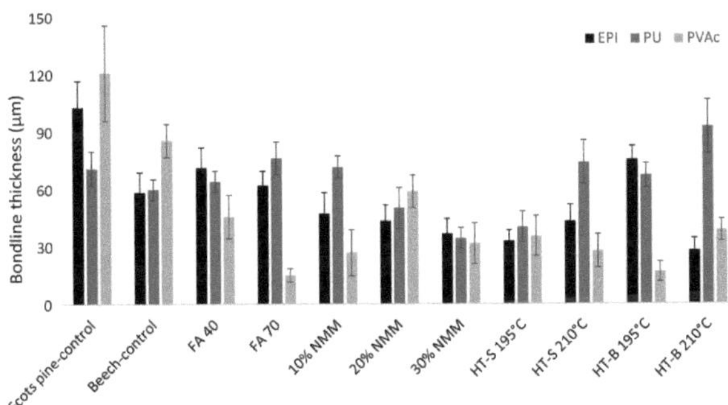

Figure 2. Mean bondline thickness values measured for furfurylated (FA 40 and 70) and NMM modified Scots pine (10%, 20% and 30% NMM), and thermally modified Scots pine (HT-S) and beech (HT-B). Error bars represent 95% confidence intervals.

modification of Scots pine and beech at 210°C. No significant change had been detected for EP of EPI following the modifications. As illustrated in Figure 2, the bondline thickness of EPI adhesive was reduced in furfurylated wood as compared to the untreated wood. It was even much lower in NMM and thermally modified Scots pine with minimum average values observed for 30% NMM and thermally modified at 195°C. Instead, the beech samples treated at 195°C showed a higher mean bondline thickness than the controls. The bondline thickness of PU adhesive was almost unchanged in the furfurylated and 10% NMM modified samples as compared to controls. It was decreased in the 20% and 30% NMM modified ones and in the Scots pine samples treated at 195°C while it was increased in the thermally modified beech samples at 210°C. The bondline thickness of PVAc adhesive was much lower in all modified Scots pine samples than the controls. It was also very low in the thermally modified beech, especially for samples treated at 195°C.

Simple correlation analysis showed that bondline thickness and EP of adhesives have no significant effect on the shear bond strength of the different modified wood materials. An exception was noted for the significant negative relationship ($r = 0.776$, .01 level) between EP and shear strength of thermally modified beech at 210° glued with EPI, and the significant positive relationship ($r = 0.692$, .01 level) between bondline thickness and shear strength of furfurylated (FA70) wood bonded with PU. It can be thus assumed that the inferior bonding of modified wood is related more to other parameters than the adhesive penetration. Such parameters can be the specific adhesive system used, the reduced chemical bonding or mechanical interlocking of adhesives, and the reduced strength of the brittle modified wood substrate.

Conclusions

This study investigated the bonding strength of different modified wood materials using two waterborne (EPI, PVAc) adhesives and one moisture curing (PU) adhesive, leading to the following conclusions:

- For all adhesives used, the shear strength significantly decreased after furfurylation and NMM modification of Scots pine samples regardless the treatment intensity. Both types of modified wood could be bonded better with PVAc adhesive while the worst bonds were created for furfurylated wood and PU adhesive. The reduction in bonding strength of furfurylated and NMM modified samples should be attributed mainly to their brittleness and inferior strength.
- Bonding strength of Scots pine and beech was also negatively affected by thermal modification. PU adhesive could provide better bonding of thermally modified beech. In all other cases, the bondline was the weakest link compared to the wood itself.
- No meaningful relationship could be established between the penetration of the adhesives, as expressed by bondline thickness and effective penetration, and the bonding strength of the different modified wood materials.

Disclosure statement

No potential conflict of interest was reported by the authors.

References

Adamopoulos, S., Bastani, A., Gascon-Garrido, P., Militz, H. and Mai, C. (2012) Adhesive bonding of beech wood modified with a phenol formaldehyde compound. *European Journal of Wood and Wood Products*, 70 (6), 897–901.

Bastani, A., Adamopoulos, S., Koddenberg, T. and Militz, H. (2015a) Study of adhesive bondlines in modified wood with fluorescence microscopy and X-ray micro-computed tomography. Submitted to *International Journal of Adhesion and Adhesives*.

Bastani, A., Adamopoulos, S. and Militz, H. (2015b) Water uptake and wetting behaviour of furfurylated, N-methylol melamine modified and heat-treated wood. *European Journal of Wood and Wood Products*. doi:10.1007/s00107-015-0919-8.

Boonstra, M. J., Tjeerdsma, B. F. and Groeneveld, H. A. C. (1998) *Thermal Modification of Non-durable Wood Species. Part 1, the Plato Technology: Thermal Modification of Wood*, Document No. IRG/WP98-40123 (Stockholm, Sweden: International Research Group on Wood Preservation).

Brady, D. E. and Kamke, F. A. (1988) Effect of hot-pressing parameters on resin penetration. *Forest Product Journal*, 38 (11/12), 63–68.

Brandon, R., Ibach, R. E. and Frihart, C. R. (2005) Effects of chemically modified wood on bond durability. In C. R. Frihart (ed.) *Wood Adhesives 2005* (San Diego, CA: Forest Products Research Society), pp. 111–114.

Chandler, J. G., Brandon, L. R. and Frihart, C. R. (2005) Examination of adhesive penetration in modified wood using fluorescence microscopy. In *ASC Spring 2005 Convention and Exposition*, 17–20 April (Columbus, OH), p. 10. Available at: http://www.treesearch.fs.fed.us/pubs/23115.

European Standard (1992) *EN 302-1: Adhesives for Load Bearing Timber Structures-Test Methods – Part 1: Determination of Bond Strength in Longitudinal Tensile Shear* (Brussels, Belgium: European Committee for Standardization).

European Standard EN 301 (2006) *Adhesives, Phenolic and Aminoplastic, for Load-Bearing Timber Structures – Classification and Performance Requirements* (Brussels, Belgium: European Committee for Standardization).

Frihart, C. R., Brandon, R. and Ibach, R. E. (2004) Selectivity of bonding for modified wood. *Proceedings of the Adhesion Society*, 27 (1), 329–331.

Gavrilović-Grmuša, I., Dunky, M., Miljkovic, J. and Djiporovic-Momcilovic, M. (2012) Influence of the degree of condensation of urea-formaldehyde adhesives on the tangential penetration into beech and fir and on the shear strength of the adhesive joints. *European Journal of Wood and Wood Products*, 70 (5), 655–665.

Hill, C. A. S. (2006) *Wood Modification: Chemical, Thermal and Other Processes* (Chichester, UK: John Wiley & Sons, Ltd).

Hse, C. H. (1971) Properties of phenolic adhesives as related to bond quality in southern pine plywood. *Forest Product Journal*, 21 (1), 44–52.

Hunt, C. G., Brandon, R., Ibach, R. E. and Frihart, C. R. (2007) What does bonding to modified wood tell us about adhesion? In *Proceedings 5th COST E34 International Workshop: Bonding of Modified Wood Bled, Slovenia* (Ljubljana: Biotechnical Faculty, Department of Wood Science and Technology), pp. 47–56.

Inari, G. N., Petrissans, M. and Gerardin, P. (2007) Chemical reactivity of heat-treated wood. *Wood Science and Technology*, 41, 157–168.

Johnson, S. E. and Kamke, F. A. (1994) Characteristics of phenolformaldehyde adhesive bonds in steam injection pressed flakeboard. *Wood and Fiber Science*, 26 (2), 259–269.

Kamke, F. A. and Lee, J. N. (2007) Adhesive penetration of wood – a review. *Wood and Fiber Science*, 39 (2), 205–220.

Kielmann, B. C., Adamopoulos, S., Militz, H. and Mai, C. (2013) Strength changes in ash, beech and maple wood modified with a n-methylol melamine compound and a metal complex dye. *Wood Research*, 58 (3), 343–350.

Kielmann, B. C., Adamopoulos, S., Militz, H., Koch, G. and Mai, C. (2014) Modification of three hardwoods with an N-methylol melamine compound and a metal-complex dye. *Wood Science and Technology*, 48, 123–136.

Kurt, R., Krause, A., Militz, H. and Mai, C. (2008) Hydroxymethylated resorcinol (HMR) priming agent for improved bondability of wax-treated wood. *Holz als Roh- und Werkstoff*, 66, 333–338.

Mahnert, K. C., Adamopoulos, S., Koch, G. and Militz, H. (2013) Topochemistry of heat-treated and N-methylol melamine-modified wood of koto (*Pterygota macrocarpa* K. Schum.) and limba (*Terminalia superba* Engl. et. Diels). *Holzforschung*, 67 (2), 137–146.

Marra, A. A. (1992) *Technology of Wood Bonding* (New York: Van Nostrand Reinhold), pp. 35–54.

Modzel, G., Kamke, F. A. and De Carlo, F. (2011) Comparative analysis of a wood-adhesive bondline. *Wood Science and Technology*, 45, 147–158.

Sahin Kol, H., Özbay, G. and Altun, S. (2009) Shear strength of heat-treated tali (*Erythrophleum ivorense*) and iroko (*Chlorophora excelsa*) woods, bonded with various adhesives. *BioResources*, 4 (4), 1545–1554.

Sernek, M., Resnik, J. and Kamke, F. A. (1999) Penetration of liquid ureaformaldehyde adhesive into beech wood. *Wood and Fiber Science*, 31 (1), 41–48.

Šernek, M., Humar, M., Kumer, M. and Pohleven, F. (2007) Bonding of thermally modified spruce with PF and UF adhesives. *Proceedings of the 5th COST E34 International Workshop, Bled, Slovenia* (Ljubljana : Biotechnical Faculty, Department of Wood Science and Technology), pp. 31–38.

Sernek, M., Boonstra, M., Pizzi, A., Despres, A. and Gérardin, P. (2008) Bonding performance of heat treated wood with structural adhesives. *Holz als Roh- und Werkstoff*, 66, 173–180.

Stamm, A. J. (1977) Dimensional stabilization of wood with furfuryl alcohol. In I. Goldstein (ed.) *Wood Chemical Aspects*, ASC Symposium Series 43 (Washington, DC: American Chemical Society), pp. 141–149.

Stig Lande, S., Westin, M. and Schneider, M. (2004) Properties of furfurylated wood. *Scandinavian Journal of Forest Research*, 19 (5), 22–30.

Suchsland, O. (1958) Über das Eindringen des Leimes bei der Holzverleimung und die Bedeutung der Eindringtiefe für die Fugenfestigkeit. *Holz als Roh- und Werkstoff*, 16, 101–108.

Vick, C. B. and Rowell, R. M. (1990) Adhesive bonding of acetylated wood. *International Journal of Adhesion and Adhesives*, 10 (4), 263–272.

van der Zee, M. E., Schipholt, N. L., Tjeerdsma, B. F., Brynildesn, P. and Mohoric, I. (2007) Glueability and paintability of furfurylated wood (Kebony). In C. A. S. Hill (ed.) *Proceedings Third European Conference on Wood Modification* (Cardiff, UK: University of Wales Bangor Biocomposites Centre), pp. 231–234.

Paper 6

The Journal of Adhesion

ISSN: 0021-8464 (Print) 1545-5823 (Online) Journal homepage: http://www.tandfonline.com/loi/gadh20

Effect of open assembly time and equilibrium moisture content on the penetration of polyurethane adhesive into thermally modified wood

Alireza Bastani, Stergios Adamopoulos & Holger Militz

To cite this article: Alireza Bastani, Stergios Adamopoulos & Holger Militz (2015): Effect of open assembly time and equilibrium moisture content on the penetration of polyurethane adhesive into thermally modified wood, The Journal of Adhesion, DOI: 10.1080/00218464.2015.1118621

Accepted author version posted online: 17 Dec 2015.

Effect of open assembly time and equilibrium moisture content on the penetration of polyurethane adhesive into thermally modified wood

Alireza Bastani[1*], Stergios Adamopoulos[2], Holger Militz[1]

[1]Wood Biology and Wood Products, Burckhardt Institute, Georg-August-University Göttingen, Büsgenweg 4, 37077 Göttingen, Germany

[2] Department of Forestry and Wood Technology, Linnaeus University, Lückligs plats 1, 351 95 Växjo, Sweden

*Corresponding author: Phone: +49-551-3933664 Fax: +49-551-399646 E-mail: abastan@gwdg.de

Received 14 August 2015; revised 07 November 2015; accepted 07 November 2015.

Abstract

The effect of wood moisture content and open assembly time on penetration of polyurethane (PU) adhesive into thermally treated Scots pine (195 and 210°C) was investigated according to effective (EP) and maximum penetration (MP) measurements using fluorescence microscopy. For samples treated at 195°C, a higher EP was noted at 8.6% equilibrium moisture content (EMC) after both assembly times (15 and 30 min) while for samples treated at 210°C, increasing wood moisture content resulted in a significant decrease in EP at 12.5% EMC after 15 min assembly time. Extending open assembly time was found to increase the EP of PU adhesive only in the case of samples treated at 195°C and with 8.6% EMC. For samples treated at both treatment temperatures and after shorter open assembly time, the highest MP observed at moderate EMC levels of 8.6 and 8.2% and the lowest at the higher EMC levels of 13.2 and 12.5%.

Keywords: Thermal modification, wood, adhesive bondlines, adhesive penetration, polyurethanes, fluorescence microscopy, bonding variables

1 Introduction

Wood modification encompasses a set of environmental friendly technologies for improving specific wood attributes such as hygroscophibity, dimensional stability, UV resistance, hardness and durability [1, 2]. However, there are still many fundamental and technological challenges to be explored in wood modification on the way to practical applications [3].

Thermally modified wood is known as one of the highly demanded materials for interior and exterior non-structural applications such as cladding, decking, flooring and furniture [4]. In addition to the positive changes also gained by other wood modifications, wood undergoes a pleasant coloration after thermal modification making it preferred for decorative purposes [5]. Industrial thermal modifications generally take place between 160°C and 260°C and change the hydrophilic properties of the treated wood due to the thermal degradation of cell wall polymers, mainly hemicelluloses [6-9]. Many commercial methods for thermal modification exist depending on the different temperatures and pressure conditions applied, and with or without oil [10]. The treatment atmosphere and other parameters are critical variables, and thermally induced chemical changes in the wood differ between various processes [11].

Decline of reactive free -OH groups of cell wall polymers by thermal modification of wood might affect its gluing or painting [12-14]. Among the different factors involved in bonding, penetration of adhesive into the wood substrate is known to affect the bond quality, and hence the performance of glued wooden structures [15-19]. Penetration is defined as the movement of liquid adhesive into the porous and complex wood tissues [20] , and can vary depending on specific wood (e.g. anatomy, moisture content, surface energy), adhesive (e.g. viscosity, pH) and process (e.g. open assembly time, temperature, pressure) related parameters [21]. Among these factors, moisture content of wood and open assembly time (time interval between spreading the adhesive on the adherent and the completion of assembly process) alter penetration by controlling the viscosity of adhesive [22], and consequently the geometry of interphase region (the volume including both wood cells and adhesive) of the adhesive bond [23, 24]. Inappropriate moisture content of wood causes bonding problems, and the attempt is to bring the moisture content of the bonded assembly close to the equilibrium moisture content of the place where the final product is going to be used [20, 25]. In most of cases, the optimal penetration and curing of the adhesive can be achieved between 4 and 10% wood moisture content [21]. Increasing the moisture content improves penetration as a result of decreased adhesive viscosity [24], and therefore bonding strength is improved [15]. Selection of a proper open assembly time is also important, because a short time does not let the adhesive to penetrate sufficiently, while a long time can cause drying of the adhesive on the surface. Bonding performance is affected negatively in both cases [26]. Moreover, the duration of open assembly time determines the production time and accordingly influences the production costs [25].

While basic knowledge of the bonding behavior of modified wood has already been established, adhesive penetration has been little studied as an influencing parameter [27-29]. Recently, it was investigated the radial penetration of three conventional cold-set wood adhesives (emulsion polymer isocyanate, polyvinyl acetate, and polyurethane) into furfurylated, N-methylol melamine and thermally modified wood [30]. The less functional -OH groups in the thermally modified wood available for chemical bonding with the reactive polyurethane adhesive than in the melamine and furfurylated wood led to a higher penetration in thermally modified wood. Polyurethane also penetrated much deeper into thermally modified wood than the two other adhesives used. This paper further reports on the penetration of polyurethane adhesive into thermally modified wood by studying the effect of two important bonding variables, wood moisture content and open assembly time.

2 Materials and methods

2.1 Thermal modification

Thermal modification was performed by Timura Holzmanufaktur GmbH (Nordhausen, Germany) through a vacuum-press dewatering procedure (Vacu³) at 195°C and 210°C with Scots pine (*Pinus sylvestris* L.) boards of primary dimensions $140 \times 10 \times 3$ cm^3 (L × W × T). The applied vacuum (150 mbar) decreased the boiling point of water resulting in an accelerated drying of the wood and the pressure was employed from the top of the boards to reduce deformation. The heat was transferred directly to the both surfaces of the boards by means of hotplates providing a uniform heat transmission, while the applied high vacuum caused the condensed moisture to be taken out of the system via dryer's wall [31]. However, some of the process details of the Vacu^3 method, such as the duration of the vacuum-press dewatering phase and the whole process length, cannot be mentioned as they are the proprietary of the producer.

2.2 Gluing and adhesive penetration

Samples with dimensions $330 \times 100 \times 6$ mm^3 (longitudinal $\times$ tangential $\times$ radial) were cut from various positions of the modified boards by paying attention to avoid defects and minor cracking. The samples were planned and then were conditioned at 55%, 78% and 94% relative humidity (RH). These three levels of RH were selected to capture the range of wood moisture content suggested for bonding with one-component moisture curing polyurethane adhesive (Jowat AG, Detmold, Germany). Conditioning of samples was possible with the use of closed vessels in which prescribed saturated salt solutions of calcium nitrate (55% RH), sodium chloride (78% RH) and lead (II) nitrate (94% RH) were placed at $20\pm1°C$. Equilibrium moisture content (EMC) calculations were based on the weight under standard conditions and the dry weight of samples. After the desired EMC was achieved, the samples were slightly planned to provide fresh tangential surfaces for gluing. The one-component polyurethane (PU) adhesive used for this study was provided by Jowat AG, (Detmold, Germany-Table 1). It is a pre-polymerized, isocyanate- functionalized polymer, which cures by the moisture content of the wood and can enter into cell lumens but hardly into cell walls. The reactive isocyanate component of the adhesive reacts with the -OH groups of the cell wall polymers leading to formation of CO_2 gas and to curing of the adhesive. The PU adhesive was applied on the tangential surfaces of the wood samples and the glued assemblies were put under a hydraulic press at 1 N mm^{-2} for 4 h at room temperature after 15 min and 30 min open assembly times as suggested by the adhesive manufacturer. Ten assemblies were prepared for each treatment temperature (195°C and 210°C), open assembly time (15 min and 30 min), and EMC level obtained after conditioning at different RH, thus 120 assemblies in total.

For microscopy, 20-30 μm thick sections exposing a bondline with a transverse surface at various positions of the assemblies were cut using a Reichert-Jung sliding microtome and stained with 0.5 % safranin O solution. Adhesive penetration was observed using an Eclipse 50i fluorescence microscope equipped with appropriate filter sets and a Sight DS-5M-L1 digital camera, and analyzed with the image analysis software NIS-Elements F (all Nikon, Düsseldorf, Germany). Following the methodology found in Sernek et al. [26], a random pattern of penetrated areas was used to measure effective penetration (EP) and maximum penetration (MP) of PU adhesive into the wood structure. EP is described as the whole adhesive parts detected in the interphase region divided by the bondline width, and MP is defined as the mean value of five deepest detected adhesive objects in this region. The used software is able to distinguish between the bright fluorescent adhesive parts and the dark background, and hence to calculate the desired parameters. For each case, ten regions of interest with dimensions $1100 \times 600 \ \mu m^2$ (width $\times$ height) were chosen to measure EP and MP.

3 Results and discussion

For the thermally modified samples at 195°C, conditioning at 55%, 78% and 94% RH yielded 5.6±0.14%, 8.6±0.21% and 13.2±0.51% EMC, respectively. The respective EMC values at each standard condition for the samples treated at the higher temperature of 210°C were 5.1±0.11%, 8.2±0.10% and 12.5±0.24%. These EMC values are the averages and standard deviations of the

EMC determinations of samples modified at the two treatment temperatures (195°C and 210°C), and will be used thereafter to present and discuss the results on PU penetration with varying EMC and open assembly time. The obtained low EMC values at each standard climate (RH and 20°C) were to be expected due to the decomposition of the hemicelluloses caused by the thermal modification [4, 32, 33]. Also, the slight decrease in EMC at each standard condition with increasing modification temperature from 195°C to 210°C was an expected result [34-36].

It should be noted that it was not aimed at the present study to provide comparisons of adhesive penetration between thermally modified Scots pine and unmodified wood. Information on the EP and MP of PU adhesive in the radial direction of thermally modified wood at 195°C and 210°C and untreated wood can be found in our previous studies without [30] and with application of pressure [37]. For the thermal modification at 195°C, a significant higher EP of PU adhesive was noted at the moderate EMC of 8.6% after both open assembly times as compared with the lower (5.6%) and higher EMC (13.2%) according to ANOVA and Tukey HSD test at P=5% (Table 2). Brady and Kamke [24] also reported a higher penetration of phenol formaldehyde adhesive into aspen wood at 15% than 4% EMC. It can be argued that a higher level of moisture content increases penetration by providing a lower viscosity of the adhesive, which leads to a better mobility of the resin solids and thus to an enchased flow. Also, adhesive viscosity can be affected at higher moisture levels by the larger number of -OH sites involved in hydrogen bonding with water molecules in comparison to the lower moisture levels. A higher penetration at elevated moisture content was also recorded for polymeric diphenylmethane diisocyanate adhesive while little penetration and absence of bond formation was observed at about 0% moisture content [15]. At this moderate EMC level (8.6%), the increase in EP was significantly higher in the case of 30 min open assembly time than 15 min (t-test at P=5%). EP attained its maximum at 8.6% EMC after 30 min open assembly time (Fig. 1c). No significant differences in EP were observed between the lower and higher EMC, and also the open assembly time had no effect on EP at these EMC levels.

The significantly higher EP of PU adhesive at the moderate EMC for the treatment at 195°C was accompanied with the highest amount and depth of penetration, especially for the shorter open assembly time of 15 min (Fig.1a). MP was much higher at the lower 5.6% EMC than the higher EMC which also represented the lowest EP at this assembly time (Fig.1b). MP values were similar among the different EMC levels after 30 min assembly time (Table 2). The maximum penetration of urea formaldehyde adhesive into beech wood was detected at 9% EMC among the different EMC steps ranged between 4-13% [26].

An improved penetration of PU adhesive into thermally treated Scots pine with increasing treatment temperature was reported previously [30, 37]. For samples treated at 210°C, EMC affected EP of PU adhesive only for the shorter open assembly time of 15 min (Table 2). For this assembly time, EP showed the lowest value at the higher EMC (12.5%) but it was significantly different only with the EP of the moderate EMC (ANOVA and Tukey HSD test at P=5%). For all EMC levels, open assembly time played no role on EP (t-test at P=5%).

Similarly to the samples treated at 195°C, the deepest penetration after 15 min open assembly time was noted at the moderate 8.2% EMC and the lowest MP at the higher 12.5% EMC (Table 2, Fig. 1d and e). After 30 min open assembly time, MP was found to decrease with increasing EMC. As discussed previously, it seems that increasing the EMC of the wood from the lower to the moderate level can facilitate adhesive penetration by developing its fluidity and movement even for a non-waterborne glue such as PU. A further increase from the moderate to the higher EMC level increases the chemical reactions between the moisture curing PU adhesive and -OH content of the wood at a certain point, which enhances the curing rate of the adhesive and hence decreases its penetration into the wood structure.

Softwoods such as pine have a simple and regular wood structure and, therefore, present an easily determinable penetration [38]. The PU adhesive used in this study is a pre-polymerized adhesive, which can enter into the cell lumens rather than penetrate into the cell walls. It is infiltrated into the wood structure mainly through the large pits of axial tracheids and ray cells as shown in a previous study [37]. Observation of microscopic photomicrographs indicated a continuous bondline containing voids in some parts with an average thickness of 30-40 μm in all cases with exception of the thick bondline detected in samples at the higher EMC, especially after 30 min assembly time (Fig. 1). The penetration happened as fully and partly field tracheids into both sides of the adherents but at the higher EMC mostly in the application side of the adhesive. It should be noted that the EP and MP data presented in this study cannot directly explain the bonding behavior of the thermally modified Scots pine since adhesive penetration is only one influencing factor on bonding, but as reported in several studies, the higher and deeper the adhesive penetration the more stronger the final bond quality [16, 19, 39, 40].

4 Conclusions

The overall results suggested a moderate EMC level of 8.6% being the optimum for an affective penetration of PU adhesive in thermally modified Scots pine treated at 195°C. That was true for both assembly times, and especially for the longer one (30 min) showing the statistically significant deepest penetration. For a higher treatment temperature (210°C) it is preferably to avoid a higher EMC level (12.5%) as it has been shown to reduce penetration of PU at a short open assembly time (15 min). With the exception of samples treated at 195°C and with moderate EMC (8.6%), increasing the open assembly time from 15 min to 30 min does not change PU penetration significantly. Therefore, for these conditions the shorter 15 min open assembly time should be used for applying PU adhesive in thermally modified Scots pine in order to save time and reduce the production costs. For the condition 195°C and EMC 8.6%, the longer 30 min open assembly time ensures a maximum PU penetration.

References

[1]   van Acker, J., and Hill, C., Proceedings of the first European Conference on wood Modification, Ghent, Belgium, pp 1-414 (2003).

[2]   Militz, H., and Hill, C., Modification: Processes, Properties and Commercialisation, (In: Militz H, Hill C (Hg.) Wood Modification: Processes, Properties and Commercialisation. Wood Modification: Processes, Properties and Commercialisation, Göttingen, Germany, 2005). pp 1-403.

[3]   Militz, H., and Lande, S., Wood Mater Sci Eng. **4** (1-2), 23-29 (2009).

DOI 10.1080/17480270903275578

[4]   Militz, H., IRG/WP 02-40241 (2002).

[5]   Mitsui, K., Takada, H., Sugiyama, M., and Hasegawa, R., Holzforschung. **55**, 601-605 (2001).

[6]   Mayes, D., Oksanen, O., ThermoWood Handbook (Finnish Thermowood Association, Helsinki, Finland, 2003).

[7]   Boonstra, M.J., Van Acker, J., Kegel, E., and Stevens, M., Wood Sci Tech. **41**, 31-57 (2007).

[8]   Esteves, B.M., and Pereira, H.M., BioResources **41**, 370-404 (2009).

[9]   Mahnert, K.C., Adamopoulos, S., Koch, G., and Militz, H., Holzforschung. **67**, 137-146 (2013).

[10]   Militz, H., Altgen, M., Processes and properties of thermally modified wood manufactured in Europe. Deterioration and Protection of Sustainable Biomaterials, ACS Symposium Series 1158:269-285 (2014).

[11]   Tjeerdsma, B.F., Stevens, M., Militz, H., and Van Acker, J., Holzforschung. **54**, 94-99 (2002).

[12]   Kamdem, D.P., Pizzi, A., and Jermannaud, A., Holz Roh Werkst. **60**, 1–6 (2002). *DOI 10.1007/s00107-001-0261-1*

[13]   Nguila Inari, G., Petrissans, M., and Gerardin, P., Wood Sci Technol. **41** (2), 157–168 (2007).

[14]   Li, X., Cai, Z., Mou ,Q., Wu, Y., and Liu, Yuan., Advanced Materials Research. **197-198**, 90-95 (2011).

[15]   Gruver, T.M., and Brown, N.R., Biores. **1**(2), 233-247 (2006).

[16]   Kamke, F.A., and Lee, J.N., Wood Fiber Sci. **39** (2), 205-220 (2007).

[17]   Guan, M., Yong, C., and Wang, L., J Appl Polym Sci **130** (2), 1345–1350 (2013).

[18]   Guan, M., Yong, C., and Wang, L., BioResources. **9** (2), 1953-1963 (2014).

[19]   Nicoli, E., Dillard, D.A., Frazier, C.E., and Zink-Sharp, A., Holzforschung. **66**, 623-631 (2012).

[20]   Marra, A.A., Technology of wood bonding, (Van Nostrand Reinhold, New York, NY, 1992). pp 35-54.

[21]   Frihart, C.R., Wood adhesion and adhesives, (In: Rowell RM (ed) Handbook of wood chemistry and wood composites, Boca Raton, FL., 2005). CRC Press LLC:215-278

[22]   Scheikl, M., Properties of the Glue Line-Microstructure of the Glue Line, (In COST Action E13: Wood Adhesion and Glued Products, Working Group 1: Wood Adhesives, State of the Art Report, 2002). pp 111-119

[23]   Smith, LA., Resin penetration of wood cell walls-implications for adhesion of polymers to wood. Ph.D. Thesis, Syracuse Univ., U.S.A (1971).

[24]   Brady, D.E., and Kamke, F.A., Forest Prod. J. **38** (11/12), 63-68 (1988).

[25]   Vallée, T., Tannert, T., and Fecht, S., J Adhes. *DOI:10.1080/00218464.2015.1071255*

[26]   Sernek, M., Resnik, J., and Kamke, F.A., Wood Fiber Sci. **31** (1), 41-48 (1999).

[27]   Sernek, M., Comparative analysis of inactivated wood surfaces. PhD dissertation, Virginia Polytechnic Institute and State University, Blacksburg, VA, 179 pp (2002).

[28]   Chandler, J.G., Brandon, L.R., and Frihart, C.R., Examination of adhesive penetration in modified wood using fluorescence microscopy, (In: ASC spring 2005 convention and exposition, April 17-20, Columbus, OH, 2005). 10 p

[29]   Adamopoulos, S., Bastani, A., Gascon-Garrido, P., Militz, H., Mai, C., Eur J Wood Prod. **70** (6), 897-901 (2012).

[30]   Bastani, A., Adamopoulos, S., and Militz, H., Eur. J. Wood Prod. **73** (5), 635–642 (2015). *DOI 10.1007/s00107-015-0920-2*

[31]   Wetzig, M., Niemz, P., Sieverts, T., and Bergemann, H., Bauphysik. **34** (1), 1-10 (2012).

[32]   Engelund, E.T., Klamer, M., and Venås, T.M., IRG/WP 10-40518 (2010).

[33]   Hill, C.A.S., Ramsay, J., Keating, B., Laine, K., Rautkari, L., Hughes, M., and Constant, B., J Mater Sci. **47**, 3191–3197 (2012).

[34]   Borrega, M., and Kärenlampi, P., Hygroscopicity of heat-treated spruce wood, (In: Proc. of the Nordic Workshop in Wood Engineering, WoodTech Sweden, Skellefteå, Sweden, 2007).

[35]   Johansson, D., Heat Treatment of Solid Wood. Effects on Absorption, Strength and Colour. PhD Thesis, Lulea University of Technology (2008).

[36]   Nguyen, C.T., Wagenführ, A., Phuong, L.X., Dai, V.H., Bremer, M., and Fische, S., BioResources. **7** (4), 5355-5366 (2012).

[37]   Bastani, A., Adamopoulos, S., Koddenberg, T., and Militz, H., Study of adhesive bondlines in modified wood with fluorescence microscopy and X-ray micro-computed tomography. Submitted to International Journal of Adhesion and Adhesives (2015).

[38]   Sonnenschein, M.F., Brune, DA., and Wendt, B.L., J Appl Polym Sci. **113** (3), 1739-1744 (2009).

[39]   Johnson, S.E., and Kamke, F.A., J Adhes. **40**, 47–61 (1992).

DOI 10.1080/00218469208030470

[40]   Modzel, G., Kamke, F.A., and De Carlo, F., Wood Sci Technol. **45**, 147–158 (2011).

Table 1 Information on the polyurethane (PU) adhesive

Property	Quantity
Solids, %	99.5
Brookfield viscosity (20°C), MPa	10,500
Density, g/cm^3	1.15

Table 2 Effective (EP) and maximum penetration (MP) of PU adhesive into thermally modified Scots pine at 195°C and 210°C with varying EMC and open assembly time (15 min and 30 min)

EMC	EP (μm)[1,2]			MP (μm)	
	15 min	30 min	t	15 min	30 min
	Thermal modification at 195°C				
5.6 %	94.3±24.0 a	104.4±26.4 a	0.895 [ns]	480	511
8.6 %	131.1±32.1 b	179.5±22.3 b	3.914 *	880	579
13.2 %	73.0±28.1 a	99.9±38.0 a	1.799 [ns]	296	576
F-value	10.839 *	22.733 *			
	Thermal modification at 210°C				
5.1 %	124.0±19.8	110.9±39.7	0.932 [ns]	673	733

	ab				
8.2 %	137.4±22.6 a	134.7±41.8	0.180 ^{ns}	852	443
12.5 %	101.9±21.4 b	89.6±42.9	0.809 ^{ns}	387	316
F-value	7.086 *	2.952 ^{ns}			

[1] Mean values ± standard deviations

[2] Values followed by a different letter within a column are statistically different (ANOVA and Tukey HSD test)

* Differences statistically significant at $P = 5$ %

^{ns} Differences not statistically significant

Fig 1 Radial penetration of PU adhesive into thermally modified Scots pine at 195°C with 8.6% (**a**) and 13.2% (**b**) EMC after 15 min open assembly time and with 8.6% EMC after 30 min open assembly time (**c**), and into samples treated at 210°C with 8.2% (**d**) and 12.5% EMC (**e**) after 15 min open assembly time

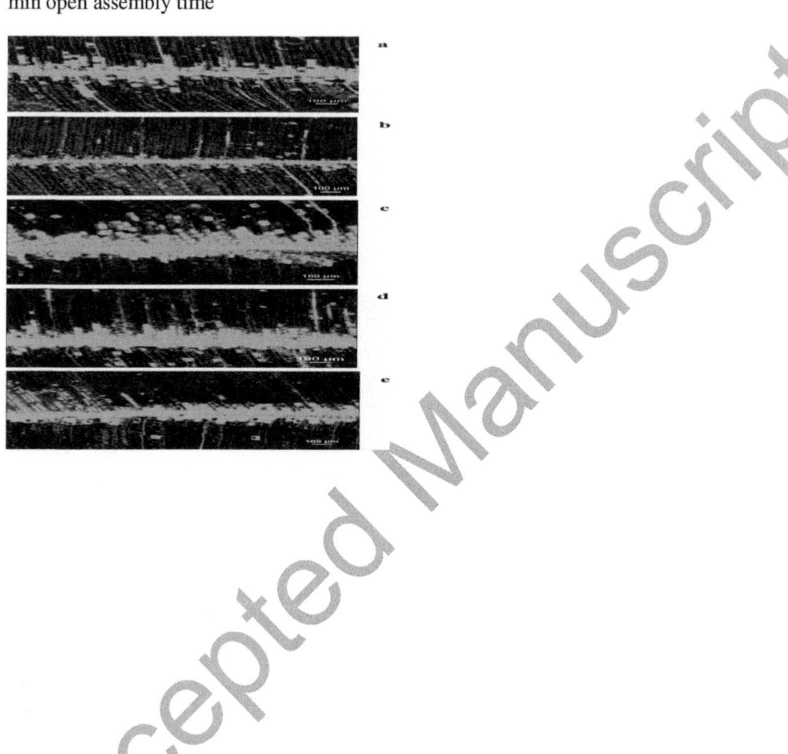

Paper 7

WOOD RESEARCH
61 (2): 2016
205-214

DEVELOPMENT OF BONDING STRENGTH OF MODIFIED BIRCH VENEERS DURING ADHESIVE CURING

Alireza Bastani, Holger Militz
Georg-August-University Göttingen, Wood Biology and Wood Products,
Burckhardt Institute
Göttingen, Germany

Stergios Adamopoulos
Linnaeus University, Department of Forestry and Wood Technology
Växjo, Sweden

Anti Rohumaa
Aalto University School of Chemical Technology
Wood Material Technology
Department of Forest Products Technology
Aalto, Finland

(Received September 2015)

ABSTRACT

This study investigated the bonding strength development of furfurylated, N-methylol melamine (NMM) modified and thermally treated birch veneers glued with hot curing phenol formaldehyde (PF) adhesive in different pressing (20, 160 s) and open assembly times (20 s, 10 min). For testing, the automated bonding evaluation system ABES was used with 2 N.mm^{-2} applied pressure at 130°C. The bonding strength of both modified and unmodified samples increased significantly by prolongation of the pressing time from 20 to 160 s in all cases and for both open assembly times. A deviation was observed for the samples treated at 220°C and at 20 s open assembly time. With the exception of NMM modified veneers, bonding strength did not change significantly by increasing the assembly time in the case of 20 s pressing for both modified and unmodified samples. At 160 s pressing time, extension of the assembly time developed a better bonding for controls, NMM modified and thermally treated veneers at 180°C. The combination of 10 min assembly time and 160 s pressing time proved as the optimal bonding condition for controls, NMM modified and thermally treated veneers at 180°C while the highest bonding strength was noted in 20 assembly time and 160 s pressing time for furfurylated veneers. In most of the cases modification affected negatively the bonding performance of the veneers, in particular for furfurylated and NMM modified samples.

KEYWORDS: Wood modification, hot curing adhesive, bonding strength, ABES, open assembly time, pressing time.

INTRODUCTION

Wood furniture and construction industries rely on effective bonding processes of wood upon cutting it into smaller pieces and rejoining these pieces together using different polymeric resins. Thus, desirable dimensions and quality of the final product can be achieved, and also the natural anisotropy of the wood can be reduced (Hass et al. 2012).

Mechanical interlocking, covalent bonding, and secondary interactions, such as Van der Waals forces and hydrogen bonding are known as involved mechanisms in adhesion between wood and adhesives while the bond formation is a dynamic process dependent on flow, transference, penetration, wetting and curing of adhesives (Marra 1992). The influencing factors on bond quality are mainly categorized in three groups: wood (e.g. species, cutting direction and free surface energy), adhesive (e.g. type, formulation, molecular weight and viscosity) and process related factors (e.g. assembly time, pressing time, pressure and temperature). These parameters have an interplay effect on each other, which also determines the final cost of production. For example, an optimized pressing or assembly time can reduce the production time and hence the production expenses (Kariž et al. 2009). Therefore, it is essential to study and optimize these parameters in order to optimise the production process and cost of glued wood products.

Several techniques have been successfully employed to study the curing of adhesives, as for example differential scanning calorimetry (Szesztay et al. 1993) and dielectrical analysis (Šernek and Kamke 2007). Evaluation of bonding strength is possible through thermomechanical analysis TMA (Soulard et al. 1999), dynamic mechanical analysis DMA (Umemura et al. 1996), torsional braid analysis TBA (Steiner and Warren 1981), integrated pressing and testing system IPATES (Heinemann 2004), and automated bonding evaluation system ABES (Humphrey 1990, 2006, Wescott et al. 2007, Jost and Šernek 2009). Among these methods, ABES provides useful information on both adhesive curing and the development of bonding strength, providing valuable data on the shear strength of the adhesive bond as a function of the pressing parameters and other conditions (Segerholm et al. 2010, Ferra et al. 2011, Rohumaa et al. 2014, Esteves et al. 2015). It should be noted that ABES is a destructive mechanical method providing only one data point per test, and thus it cannot be employed for continuous testing during the hot pressing of wood-based composites (Ferra et al. 2011).

During the last years, wood modification techniques showed a high potential to improve wood properties such as durability, weathering and UV resistance, dimensional stability, and hardness (Militz et al. 1997; Hill 2006). One of the challenges in this area is to effectively bond the different modified materials as their physical and chemical aspects are substantially altered by the passive, active (chemical), or thermal modifications in comparison to the unmodified wood (Boonstra and Tjeerdsma 2006, Rowell 2006, Nguila Inari et al. 2007). To achieve this, a thorough understanding of the various factors affecting the interaction of the adhesives and the modified adherents is needed. This study investigated the development of bonding strength of furfurylated, melamine and thermally modified birch veneers during adhesive curing of phenol formaldehyde adhesive (PF) as a function of open assembly and pressing times by using ABES.

MATERIAL AND METHODS

Wood material and modifications

Silver birch (*Betula pendula* Roth) veneers with dimensions 150 × 150 × 0.8 mm (L × W × T) were provided from Southern Finland (Rusutjärvi) and used for the modifications. Thermal modification was performed at 180 and 220°C using a lab scale heating oven (UNOX S.p.A, Italy) through 9 gradual steps of heating by circulation of hot air and steam in the system, started at 60°C up to the maximum applied temperature (180 and 220°C) followed by a cooling down phase. The duration of the main heating phase (application of maximum temperature) was 3 h in each treatment. For melamine modification of the veneers, N-methylol melamine (NMM) resin Madurit MW840/75WA (Ineos Melamines GmbH, Frankfurt, Germany) was provided as an aqueous stock solution with a solid content of approx. 75 %. The solution of 20 % NMM solid content was used for impregnation of veneers in a steel vessel with 30 m vacuum (60 mbar) followed by soaking of the veneers in the treatment solution for 2 h, then pre-drying at room temperature (1 day), and finally gradual drying/curing phase at several temperatures of 40-120°C (1 day). Furfurylation was carried out by Skog og landskap (Ås, Norway) by using an industrially known FA70 mixture of furfuryl alcohol resulting to 75 % weight percent gain. Prior to testing, all veneers were stored in a climate chamber and conditioned at 20°C and 65 % RH.

Gluing and ABES testing

The hot-curing liquid phenol-formaldehyde (PF) adhesive used for this study was provided by Dynea Chemicals (Oy, Hamina, Finland) with the commercial name Prefere 14J021 (Tab. 1). It is made up of linear or branched oligomeric molecules in an aqueous solution, which creates a three-dimensionally crosslinking structure after curing (Laborie 2002, Dunky 2004). The adhesive was applied manually on a 5 × 20 mm area of the veneer samples with dimensions 20 × 117 mm by a micropipette (HandyStep electronic®, Wertheim, Germany).

Tab. 1: Information on the PF adhesive.

Solids (%)	Brookfield viscosity (20°C), (MPa)	Density (g.cm^{-3})	pH
49	300	1.21	12

An automated bond evaluation system ABES (Incorporated, Corvallis, OR, USA) was used for measuring bonding strength (Fig. 1).

Fig. 1: Automated bond-evaluation system (ABES) used for the evaluation of bonding strength of veneers glued with PF adhesive.

After application of the adhesive on each veneer surface, the two bonded veneers were mounted in-between the holding jaws of the device at a fixed open assembly time of 20 s and

10 min (time interval between spreading the adhesive on the adherent and the completion of assembly process), and then the assembly was pressed at 2 N.mm^{-2} and 130°C for the designated period of time. Finally, the two glued veneers of the assembly were separated by movement of the holding jaws and ABES provided the force, in N, necessary to break the glue line. The bonding strength was determined by the following formula:

$$Bonding\ strength\ =\ \frac{Force\ (N)}{Bonded\ area\ (mm^2)} \qquad (N.mm^{-2})$$

Four combinations of open assembly (20 s, 10 min) and pressing (20, 160 s) times were used and can be seen in Tab. 2 together with all the other variables for ABES testing. In detail, the assembly time of 20 s reflects the initial wetting properties of the veneer surface (e.g. the ability of liquid glue to provide interfacial affinity for an adherent and to flow over its surface) and 10 min assembly time is close to the industrial process.

Tab. 2: Testing parameters used to measure bonding shear strength with ABES.

Pressure (N.mm^{-2})	Amount of adhesive, (g.m^{-2})	Curing temperature	Bonded area (mm^2)	Combination of open assembly and pressing times
2	100	130°C	4 × 20	20 s / 20 s 20 s / 160 s 10 min / 20 s 10 min / 160 s

Pressing time of 20 s refers to the condition under which the tack properties of the glue (e.g. the adhesive feature that allows formation of a bond of measurable strength instantly after adhesive and adherent contact under low pressure) can be evaluated, while 160 s pressing time is close to the industrial practice. Ten (10) replicates were used for each case, e.g. type of modification and combination of open assembly and pressing time. The total number of the tested assemblies was 200 (10 replicates × 5 treatments × 4 combinations of open assembly and pressing times).

RESULTS AND DISCUSSION

Tab. 3: Bonding strength of control, N-methylolmelamine (NMM), furfurylated (FA) and thermally (TM) modified birch veneers glued with PF adhesive at 130°C by using different combinations of open assembly and pressing times[1, 2].

Combination of open assembly and pressing times	Control	NMM	FA	TM 180°C	TM 220°C
20 s / 20 s	4.02±0.3 [a]	2.17±0.1 [a]	1.54± 0.1 [a]	3.55±0.2 [a]	4.38±0.4 [ab]
20 s / 160 s	6.55±0.8 [b]	5.12±0.5 [b]	8.64±0.6 [b]	5.91±0.6 [b]	5.07±0.5 [ab]
10 min / 20 s	3.77±0.5 [a]	1.37±0.1 [c]	1.90±0.2 [a]	3.65±0.3 [a]	4.24±0.7 [a]
10 min / 160 s	9.12±1.3 [c]	6.22±0.7 [d]	7.56±0.8 [c]	8.25±0.9 [c]	5.85±0.8 [b]
F	60.408*	170.621*	349.820*	102.750*	9.475*

[1]Mean values ± standard deviations
[2]Values followed by a different letter within a column are statistically different (ANOVA and Tukey HSD test)
* Differences statistically significant at P = 5 %.

The shear strength bonding results with the relevant statistical analysis for differences according to ANOVA and Tukey HSD test for P=5 % are presented in Tab. 3. For control samples, extending of the pressing time from 20 to 160 s significantly increased the shear strength values in both open assembly times. Jost and Šernek (2009) also reported a dependency between the curing of PF adhesive and pressing time when bonding beech veneers at 160°C. Open assembly time was found to affect bonding strength only for the high pressing time of 160 s as its prolongation from 20 s to 10 min provided a significant higher bonding strength. Actually, bonding strength achieved its peak for the higher assembly and pressing times (10 min and 160 s) and it was significantly different as compared to any other combination.

The various modifications followed more or less the trends of control samples in bonding strength development with changing open assembly and pressing times (Tab. 3). There was only one exception for the positive effect on bonding strength of the increased pressing time at each open assembly time, and it was noted in the case of thermal modification at 220°C. Although the bonding strength at 20 s open assembly time increased numerically with extending the pressing time from 20 to 160 s, the difference was not statistically significant. With the exception of the decreased bonding strength of FA modified veneers and also the thermal modification at 220°C for which open assembly time had no effect on the development of bonding strength for both pressing times, in other cases the increase in open assembly time from 20 s to 10 min at the high pressing time of 160 s produced significant higher bonding strength. Irrespective of NMM modified veneers, bonding strength remained unchanged at the low pressing time of 20 s with increasing the open assembly time in all other modifications as also noted for the controls. At this pressing time, the increase of open assembly time from 20 s to 10 min resulted to a significantly lower bonding strength of NMM modified veneers. Šernek et al. (1999) reported on little effect of extending open assembly time on penetration of urea-formaldehyde (UF) adhesive into beech.

Another similarity of the different modified samples to controls was the significant highest bonding strength value for the combination of the prolonged open assembly and pressing times (10 min and 160 s). However, for thermally modified veneers at 220°C, this obtained value was only numerically (not statistically) higher than the rest of the cases while for the FA modified veneers the combination of 20 s open assembly time and 160 s pressing time gave the best performance. The curing behaviour of four synthesized UF resins was evaluated using ABES machine by considering the effect of pressing parameters (e.g. temperature, adhesive and hardener ratios) on shear strength during hot pressing. The results revealed a different bonding process for each type of the used resins with the best values of shear strength for the resin produced by alkaline–acid process at 100°C press temperature and 80 s press time (Ferra et al. 2011).

Changes in the hygroscopic behaviour and wetting properties of wood caused by various modifications can potentially affect and hinder proper curing of adhesives and hence the bonding performance (Boonstra et al. 1998, Petrissans et al. 2003, Šernek et al. 2008, Bastani et al. 2015a, b). As expected, for the majority of combinations of open assembly and pressing times the modifications had a negative effect on the bonding strength in comparison to controls, especially for furfurylation and NMM modification (Tab. 3). The bonding of these two types of modified veneers with such a hot curing PF adhesive is more dependent on pressing time than any other types of wood used in this study, as it was proved by their low bonding strength at 20 s pressing time in both assembly times, mainly due to poor initial wetting of the veneers surface after modification. The chemical combination of the surface of these modified materials can change by extending pressing time which resulted in an obviously much higher bonding strength values after 160 s of pressing. The bonding property of the same types of modified wood materials glued with three common cold-set wood adhesives together with details on the relationship between bonding strength and bondline thickness, and adhesive penetration was investigated in previous studies (Bastani et al. 2015c, Bastani et al. 2016). It was found that the reduction in bonding strength of furfurylated and NMM modified wood with emulsion polymer isocyanate (EPI), polyvinyl acetate (PVAc) and polyurethane (PU) adhesives should be attributed mainly to the lower strength of the brittle modified wood and to the reduced internal surfaces for chemical bonding or mechanical of adhesives in the bulked modified wood. In the case of thermally modified Scots pine wood bonded with PU, it was found that the bondline was the weakest link compared to the wood itself. It was thus implied that the decreased adhesion should be attributed to the existence of less polar groups for bonding in thermally modified wood and to its poorer wettability hindering the proper curing of PU.

An overview on the obtained results of present study for each combination of pressing and assembly times indicated that in 20 s pressing and assembly times a decreased adhesion of about 62, 46 and 12 % was noted for FA, NMM and thermally modified veneers (180°C), respectively. Samples treated at 220°C showed 8 % higher bonding strength than controls for the above mentioned combination. In the case of 20 s assembly time and 160 s pressing, the bonding strength was also reduced for all types of modified wood, with the exception of 24 % higher bonding of FA modified veneers as compared to controls. For combination of 10 min assembly time and 20 s pressing, the bonding strength reduced after NMM modification (64 %) and furfurylation (50 %) while thermally treated veneers at 220°C represented the highest values among all cases, even 11 % higher than those of controls. The bonding strength was also reduced for all types of modified woods in the case of 10 min assembly time and 160 s pressing, 36 and 32 % for thermally treated (220°C) and NMM modified samples and 17 and 10 % for furfurylated and thermally treated samples at 180°C as compared to controls. The decreased adhesion properties of the modified veneers could be attributed to the poorer wettabilty of wood surface after modification as proved by contact angle and surface energy measurements on the same modified materials Bastani et al. (2015a). The reduction in shear strength after modification was also noted for samples bonded with differnt thermoplastic and thermosetting adhesives (Vick and Rowell 1990), with PF and UF (Šernek et al. 2007) and melamine-urea-formaldehyde adhesive (Šernek et al. 2008). A noticeable finding in this research was the highly improved bonding strength of furfurylated veneers by increasing pressing time in both assembly times, five and a half times higher in the case of 20 s assembly time and about four times higher in 10 min assembly time which illustrates the important role of temperature in effective bonding of FA modified veneers with hot curing PF adhesive. Furfuryl alcohol is known to react strongly with itself and starts to polymerize at presence of heat to build a cross-linked polymer, which might help to improve the bonding strength of furfurylated assemblies (Larsson-Brelid 2013). However, the curing chemistry of PF glue is a complex matter and dependent on several influencing factors (Laborie 2002, Dunky 2004).

CONCLUSIONS

Bonding strength of different types of modified birch veneers bonded with hot curing PF adhesive was examined in different pressing and open assembly times. The obtained results lead to the following conclusions:

- With the exception of modified veneers at 220°C and at 20 s open assembly time, extending of the pressing time from 20 to 160 s significantly improved the bonding strength in all cases for untreated and treated samples in both open assembly times.

- With the exception of NMM modified veneers, for both modified and control samples, no significant change was seen in bonding strength by increasing the assembly time for the 20 pressing while at 160 s pressing, prolongation of assembly time improved bonding of controls, NMM modified and thermally treated veneers at 180°C.
- The combination of 10 min assembly time and 160 s pressing time provided the highest bonding strength for controls, NMM modified and thermally treated veneers at 180°C while furfurylated samples gained the highest values in 20 assembly and 160 s pressing times.
- In most of the cases, and especially in furfurylated and NMM modified samples, modification showed a negative effect on bonding strength. The bonding process of these two types of modified materials seemed to be highly dependent on pressing time. The obtained results of this study can be useful for effective bonding of various modified veneers with hot curing adhesives.

ACKNOWLEDGMENT

Financial support from COST Action committee (COST-STSM-FP1303-18764) and the provision of ABES testing device by Aalto University is gratefully acknowledged.

REFERENCES

1. Bastani, A., Adamopoulos, S., Militz, H., 2015a: Water uptake and wetting behaviour of furfurylated, N-methylol melamine modified and heat-treated wood. Eur. J. Wood Prod. 73(5): 627-634.
2. Bastani, A., Adamopoulos, S., Militz, H., 2015b: Gross adhesive penetration in furfurylated, N-methylol melamine-modified and heat-treated wood examined by fluorescence microscopy. Eur. J. Wood Prod. 73(5): 635-642.
3. Bastani, A., Adamopoulos, S., Koddenberg, T., Militz, H., 2015c: Study of adhesive bondlines in modified wood with fluorescence microscopy and X-ray micro-computed tomography, under review in International Journal of Adhesion and Adhesives.
4. Bastani, A., Adamopoulos, S., Militz, H., 2016: Shear strength of furfurylated, N-methylol melamine and thermally modified wood bonded with three conventional adhesives, under review in Wood Material Science and Engineering, DOI:10.1080/17480272.2016.1164754.
5. Boonstra, M.J., Tjeerdsma, B.F., Groeneveld, H.A.C., 1998: Thermal modification of non-durable wood species. Part 1. The Plato technology: Thermal modification of wood. In: International Research Group on Wood Preservation, Document Nr. IRG/WP98-40123, Stockholm, Sweden, 13 pp.
6. Boonstra, M.J., Tjeerdsma, B.F., 2006: Chemical analysis of heat treated softwoods. Holz als Roh- und Werkstoff 64(3): 204-211.
7. Dunky, M., 2004: Adhesives based on formaldehyde condensation resins. Macromolecular Symposia 217(1): 417-430.
8. Esteves, B., Martins, J., Martins, J., Cruz-Lopes, L., Vicente, J., Domingos, I., 2015: Liquefied wood as a partial substitute of melamine-urea- formaldehyde and urea-formaldehyde resins. Maderas, Cienc.tecnol. 17(2), DOI:10.4067/S0718-221X2015005000026

9. Ferra, J.M.M., Ohlmeyer, M., Mendes, A.M., Costa, M., Carvalho, L., Magalhães, F., 2011: Evaluation of urea-formaldehyde adhesives performance by recently developed mechanical tests. International Journal of Adhesion & Adhesives 31: 127-134.

10. Hass, P., Wittel, F.K., Mendoza, M., Herrmann, H.J., Niemz, P., 2012: Adhesive penetration in beech wood: Experiments. Wood Sci. Technol. 46(1): 243-256.

11. Heinemann, C., 2004: Characterization of the curing process of adhesives in wood-particle matrices by evaluation of mechanical and chemical kinetics. Hamburg: Institute for Wood Technology, University of Hamburg.

12. Hill, C.A.S., 2006: Wood modification: Chemical, thermal and other processes. John Wiley & Sons, Chichester, UK, 239 pp.

13. Humphrey, P.E., 1990: Device for testing adhesive bonds. United States Patent 5.176.028.

14. Humphrey, P.E., 2006: Temperature and reactant injection effects on the bonding kinetics of thermosetting adhesives. In: Frihart CR Wood adhesives 2005. Forest Products Society, Madison. Pp 311-316.

15. Jost, M., Šernek, M., 2009: Shear strength development of the phenol-formaldehyde adhesive bond during cure. Wood Sci. Technol. 43(1): 153-166.

16. Kariž, M., Jošt, M., Šernek, M., 2009: Curing of phenol-formaldehyde adhesive in boards of different thicknesses. Wood Research 54(2): 41-48.

17. Laborie, M.P., 2002: Investigation of the wood/phenol-formaldehyde adhesive interphase morphology. PhD thesis, Virginia Polytechnic Institute and State University.

18. Larsson-Brelid, P., 2013: Benchmarking and state of the art for modified wood. Report 2013: 54, SP Wood Technology. SP Technical Research Institute of Sweden, 30 pp.

19. Marra, A., 1992: Technology of wood bonding principles in practice. Van Nostrand Reinhold, New York, 454 pp.

20. Militz, H., Beckers, E.P.J., Homan, W.J., 1997: Modification of solid wood: Research and practical potential. In: The International Research Group on Wood Preservation, Document Nr. IRG/WP 97-40098, Whistler, Canada.

21. Nguila Inari, G., Petrissans, M., Gerardin, P., 2007: Chemical reactivity of heat-treated wood. Wood Sci. Technol. 41(2): 157-168.

22. Petrissans, M., Gerardin, P., El Bakali, I., Serraj, M., 2003: Wettability of heat treated wood. Holzforschung 57(3): 301-307.

23. Rohumaa, A., Hunt, C.G., Frihart, C.R., Saranpää, P., Ohlmeyer, M., Hughes, M., 2014: The influence of felling season and log-soaking temperature on the wetting and phenol formaldehyde adhesive bonding characteristics of birch veneer. Holzforschung 68(8): 965-970.

24. Rowell, R., 2006: Chemical modification of wood: A short review. Wood Mater. Sci. Eng. 1(1): 29-33.

25. Segerholm, K., Wålinder, M., Holmberg, D., 2010: Adhesion studies of Scots pine-polypropylene bond using ABES. In: Proceedings of the 6th meeting of the Nordic-Baltic Network in Wood Material Science and Engineering, WSE, Tallinn, Estonia. Pp 142-146.

26. Šernek, M., Resnik, J., Kamke, F.A., 1999: Penetration of liquid ureaformaldehyde adhesive into beech wood. Wood Fiber Sci. 31(1): 41-48.

27. Šernek, M., Kamke, F.A., 2007: Application of dielectric analysis for monitoring the cure process of phenol formaldehyde adhesive. Int. J. Adhes. 27(7): 562-567.

28. Šernek, M., Humar, M., Kumer, M., Pohleven, F., 2007: Bonding of thermally modified spruce with PF and UF adhesives. In: Proceedings of the 5th COST E34 International Workshop, Bled, Slovenia. Pp 31-38.

29. Šernek, M., Boonstra, M., Pizzi, A., Despres, A., Gerardin, P., 2008: Bonding performance of heat treated wood with structural adhesives. Holz als Roh- und Werkstoff 66(3): 173-180.
30. Soulard, C., Kamoun, C., Pizzi, A., 1999: Uron and uron-urea-formaldehyde resins. J. Appl. Polym. Sci. 72(2): 277-289.
31. Steiner, P.R., Warren, S.R., 1981: Rheology of wood-adhesive cure by torsional braid analysis. Holzforschung 35(6): 273-278.
32. Szesztay, M., Laszlohedvig, Z., Kovacsovics, E., Tudos, F., 1993: DSC application for characterization of urea/formaldehyde condensates. Holz als Roh- und Werkstoff 51(5): 297-300.
33. Umemura, K., Kawai, S., Mizuno, Y., Sasaki, H., 1996: Dynamic mechanical properties of thermosetting resin adhesives II. Urea resin. Mokuzai Gakkaishi 42(5): 489-496.
34. Vick, C.B., Rowell, R.M., 1990: Adhesive bonding of acetylated wood. International Journal of Adhesion & Adhesives 10(4): 263-272.
35. Wescott, J.M., Birkeland, M.J., Traska, A.E., Frihart, C.R., Dally, B.N., 2007: New method for rapid testing of bond strength for wood adhesives. In proceedings of the 30th Annual meeting of The Adhesion Society Inc. Tampa Bay, Florida. Pp 219-221.

ALIREZA BASTANI*, HOLGER MILITZ
GEORG-AUGUST-UNIVERSITY GÖTTINGEN
WOOD BIOLOGY AND WOOD PRODUCTS
BURCKHARDT INSTITUTE
BÜSGENWEG 4
37077 GÖTTINGEN, GERMANY
PHONE: + 49-551-3933664
Corresponding author: abastan@gwdg.de

STERGIOS ADAMOPOULOS
LINNAEUS UNIVERSITY
DEPARTMENT OF FORESTRY AND WOOD TECHNOLOGY
LÜCKLIGS PLATS 1
351 95 VÄXJO
SWEDEN

ANTI ROHUMAA
AALTO UNIVERSITY SCHOOL OF CHEMICAL TECHNOLOGY
WOOD MATERIAL TECHNOLOGY
DEPARTMENT OF FOREST PRODUCTS TECHNOLOGY
P.O. BOX 16400
00076 AALTO
FINLAND

Curriculum Vitae

Personal

Name: Alireza Bastani
Date of birth: 18 September 1980
Place of the birth: Ahvaz, Iran

Education

April 2011-March 2016   Ph.D. Wood Biology and Wood Products, Göttingen University, Germany

2004-2006   M.sc, Wood and Paper Science and Technology, Noshahr Chalous Azad University, Iran

1999-2004   B.sc, Wood and Paper Science and Technology, Karaj Azad University, Iran

Employment record

April 2011-March 2016   Wood Biology and Wood Products, Georg-August-Universität Göttingen, Büsgenweg 4, 37077 Göttingen, Germany

2010-2011   Sima choob Company, manufacturer and designer of office furniture, wooden artifacts, conference halls and interior decoration, Tehran, Iran

2007-2010   Private workshop, kitchen and interior design, Karaj, Iran

2005-2006   Kian carton Company, producer of carton boxes, Tehran, Iran